Electrical and Electronic Drafting

ELECTRONIC TECHNOLOGY SERIES

VOLUME EDITOR
Irving L. Kosow

Herbert W. Richter
ELECTRICAL AND ELECTRONIC DRAFTING

Curtis Johnson
PROCESS CONTROL INSTRUMENTATION TECHNOLOGY

Luces M. Faulkenberry
AN INTRODUCTION TO OPERATIONAL AMPLIFIERS

Joseph Greenfield:
PRACTICAL DIGITAL DESIGN USING ICs

Herbert W. Richter
Kalamazoo Valley Community College

Electrical and Electronic Drafting

John Wiley & Sons, Inc.
New York
Santa Barbara
London
Sydney
Toronto

Cover & Text Designed by
Jules Perlmutter, A Good Thing, Inc.
Joan Tobin supervised production

Library of Congress Cataloging in Publication Data:

Richter, Herbert W
    Electrical and electronic drafting.

  (Electronic technology series)
  Includes indexes.
    1. Electric drafting. 2. Electronic drafting.
I. Title.

TK431.R5   604'.2'6213   76-20506
ISBN 0-471-72035-6

Printed in the United States of America

10 9 8 7 6 5 4

To my wife
Madelin
and my children
John and Anne

Preface

This book was written expressly to interpret industrial drafting practices for the beginner. It introduces the various drafting techniques and types of drawings used in the design and construction of electronic and electrical equipment. Community college students, draftsmen, engineering technicians, and others planning a career in the electronics industry will find it a practical working text of great value. This book was also planned to coordinate with the recent emphasis on career education in high schools. The material and order of its presentation are largely based on the very successful electronics drafting course taught at the Kalamazoo Valley Community College in Kalamazoo, Michigan since 1968.

Since electronics drafting is likely to be offered early in a community college electronics program, circuit theory (Chapter 5) is limited to developing the students' knowledge of basic circuit recognition. References are included at the end of each chapter for those students who wish further information on a particular subject. The symbols of common electronic and electrical components are pictorially illustrated in Chapters 4, 6, 11, and 12 so that the beginner may quickly associate the symbol with the component. All electronic symbols conform to American Standard Y32.2-1970, which has been incorporated by the Institute of Electrical and Electronic Engineers as IEEE Std. 315-1971. Graphic symbols for electrical wiring and layout diagrams used in architecture and building construction conform to the American National Standards Institute (ANSI) Y32.9-1972, also adopted by the IEEE.

This material presented here can be taught in a single-semester course, although the instructor may wish to place more emphasis on the electronic chapters and less on the architectural construction and industrial wiring chapters, depending on the employment opportunities in the area. The problems and self-evaluation questions presented in each chapter are directed toward self-study. Each problem includes working data, specific instructions, and is illustrated when necessary; however, I recommend that the instructor supplement the problems with actual *bread-boarded* circuits, manufactured assemblies, or subassemblies, since such problems will more closely relate to actual job experience. It is not necessary for the components and assemblies to be in operating condition.

Production and assembly drawing are emphasized since electronics drafting is almost entirely devoted to this area. Chapters 7 to 10 are therefore concerned with isometric and perspective pictorial drawings, which are necessary in electronic manufacturing plants for use by possibly untrained assemblers and sales and purchasing personnel.

I gratefully acknowledge the support and help of the Wiley staff, particularly Alan B. Lesure and Dr. Irving L. Kosow. Special thanks must go to manufacturers who generously contributed photographs, circuits, and other useful data, and gave permission for their use.

Kalamazoo, Michigan   Herbert W. Richter

Contents

Electrical and Electronic Drafting

Chapter 1 Instruments and Drawing Techniques

Instructional Objectives

To learn:

1. The desirable characteristics and properties of a drawing board and T-square.
2. The purposes and uses of a T-square and triangles.
3. To draw parallel horizontal lines and lines at any angle.
4. The correct techniques of using the instruments.
5. The desirable characteristics and properties of drawing instruments and papers.
6. The application of templates and other drafting aids.
7. The proper care of drawing equipment.
8. Desirable work habits.

Self-Evaluation Questions

Test your prior knowledge of the information in this chapter by answering the following questions. Watch for the answers as you read the chapter. Your final evaluation of whether you understand the material is measured by your ability to answer these questions. When you have completed the chapter, return to this section and answer the questions again.

1. Give the name of the type of drawing in which all lines lie at right angles to the plane of projection.
2. Give the advantages in fastening drawing paper to a drawing board using adhesive tape rather than thumbtacks.
3. How should the drafting pencil be held and inclined when drawing a line using a straight edge?
4. What is the minimum drafting equipment needed to produce an orthographic drawing?
5. Describe the proper method of drawing a perpendicular to a given nonhorizontal line.
6. What are the important characteristics of vellum drawing paper?
7. What method would you use to draw a line accurately connecting two points?
8. What effect does conversion to the metric system have on drawing techniques?

1

9. How could an angle of 75° be drawn using two drafting triangles and a T-square?

10. Describe your method to draw an angle of 22.5° using T-square, triangles, and compass.

1-1 Drawing Boards and Tables

The *drawing board* must be large enough to accommodate the largest drawing that may be contemplated. Its surface must be smooth flat, and constructed of a warp-free material. If a *T-square* rather than a drafting machine is used, one edge of the board must be a *straight* or reference edge for perfect alignment with the *head* of the T-square. Often to protect the board surface, a sheet of paper slightly smaller in size than the board dimensions is fastened to the surface with adhesive tape. This provides a clean, nonslip, and only slightly resilient surface that protects the board from nicks, cuts, or dents.

Many different board materials may meet these specifications: smooth pine, plywood, particle board, linoleum, or sheet metal. Boards 97 × 122 cm (38 × 48 in.) and larger may be provided with either permanent or folding legs. Such *drafting tables* are also provided with an adjustment so that the drawing surface may be made to slope at a convenient angle as shown in Figure 1-1. The overall height of most pedestal-type tables may be adjusted at the legs for a less tiring posture; desk-type drafting tables may not have this advantage but often contain storage space. The table shown in Figure 1-1 is actually a specialized drafting table in that it has a *translucent* glass or plastic working surface illuminated from below the tabletop; this construction is extremely useful for tracing work.

Figure 1-1 Pedestal-type shadowless tracing table. (Courtesy Bruning Div. AM.)

1-2 T-Squares and Triangles

The purpose of a *T-square* is to draw *horizontal* and *parallel* lines. The T-square, in conjunction with drafting *triangles,* is used to draw *vertical* lines in a fundamental drafting procedure called *orthographic* or multiview projections (Fig. 1-11). A clear plastic straight edge may be bonded to the blade providing a convenient "see through" edge.

The T-square is possibly the most easily damaged drafting tool. If it is dropped, the head may no longer be rigidly attached at right angles to the blade. Lines drawn when the head is loose are no longer parallel to each other. Obviously a nicked straight edge results in drawing a discontinuous line.

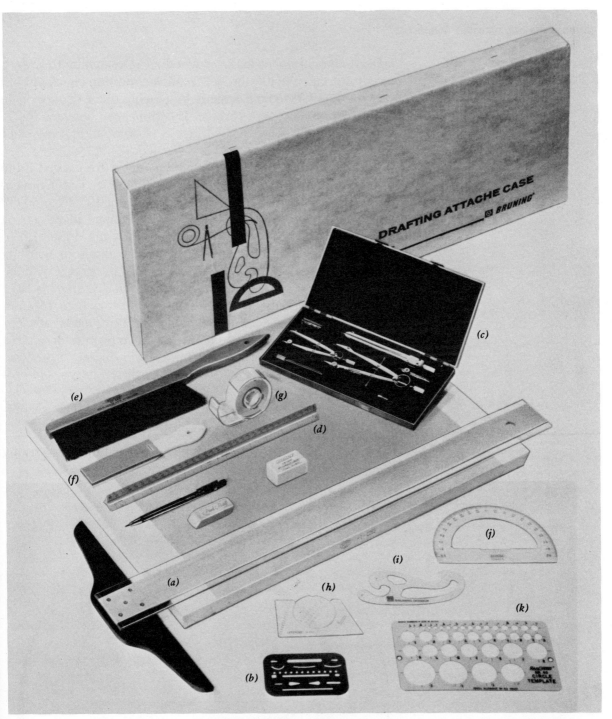

Figure 1-2 Drafting attache case. (Courtesy Bruning Div. AM.) (*a*) T-square; (*b*) erasing shield; (*c*) instruments case; (*d*) triangular scale; (*e*) dust brush; (*f*) sandpaper block; (*g*) cellophane tape; (*h*) lettering guideline template; (*i*) French curve; (*j*) protractor; (*k*) circle template.

Drafting equipment made of wood has a tendency to warp if care is not taken in its storage. A quick condition check of both the board and T-square is made by placing the T-square blade on edge upon the board surface. If light is seen between the T-square edge and the board surface it is obvious that either or both the T-square and board are warped.

A 30 to 60° and a 45° *right-angled triangle* are needed if the draftsman does not use a drafting machine (Fig. 1-5). Conventional triangles permit the accurate drafting of 30°, 45°, and 60° angles with respect to the horizontal or T-square blade. Using both the 45° and 30 to 60° triangles permits the draftsman to construct angles in multiples of 15°. For example, if the long side or *hypotenuse* of a 45° triangle is placed in contact with the short side of a 30 to 60° triangle, the total *included angle* is 105°.

A *protractor* is also useful when angles other than 15° multiples are drawn. These instruments are now made of clear plastic for "see through" convenience. The protractor is a flat semicircular or circular device engraved in degrees from zero through 180° or 360°, respectively. Three different types of protractors are shown in Figure 1-3.

When using either the T-square or triangles several basic precautions should be observed. To draw a horizontal line:

1. Press the head of the T-square firmly against the left working edge of the drawing board.

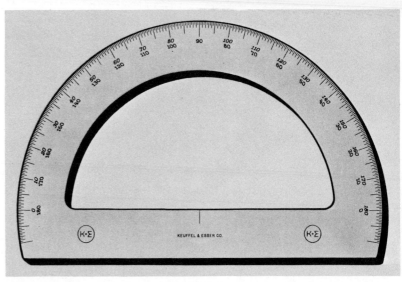

Figure 1-3 Protractors. (Courtesy of Keuffel & Esser Co.)

(a) 180° protractor

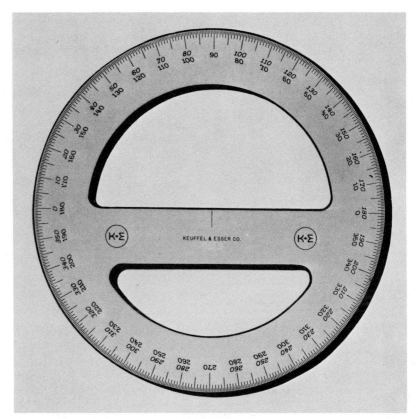

(b) 360° protractor

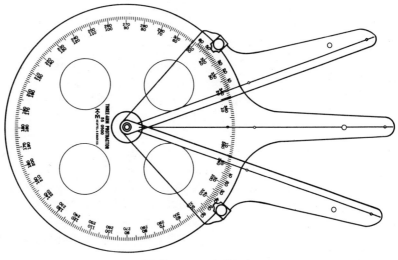

Figure 1-3 (*continued*)

(c) three-arm protractor

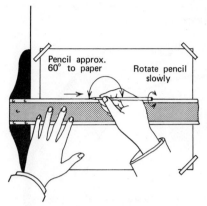

(a) Drawing a horizontal line

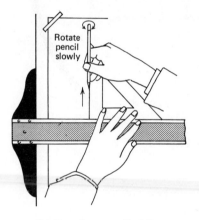

(b) Drawing a vertical line

Figure 1-4 Pencil techniques.

2. Press the T-square blade tightly against the paper.
3. Keep the pencil vertical to the board while drawing the line. The pencil may be leaned slightly in the direction of motion.
4. If the pencil is rotated slowly as the line is drawn, the pencil point wears more uniformly.
5. The pencil point should not touch the bottom edge of the T-square blade, since the blade and drawing may become soiled. This is particularly important when drawing inked lines.

To draw either vertical or inclined lines:

1. The base of a triangle should rest evenly and firmly against the T-square blade.
2. The pencil is maintained vertical to the board but the line is drawn *upward,* or away from the draftsman; the pencil may be leaned slightly in the direction of motion. See Figure 1-4.
3. When a line is to be drawn between two points, the pencil is placed vertically at one of the points. A straight edge is then moved to touch the pencil point and aligned with the second point before drawing the line.

The T-square and a triangle, or a pair of triangles, may also be used to draw a line parallel to a given non-horizontal line:

1. Move the triangle and T-square as a unit until the hypotenuse of the triangle lines up with the given line.
2. Hold the T-square firmly in position.
3. Slide the triangle away from the given line.
4. Draw the required line along the hypotenuse.

1-3 Drafting Machines *Drafting machines* may be purchased apart from or with a drawing board. The *parallel-rule mechanism,* known as a drafting machine, is shown in Figure 1-5. An L-shaped straight-edge replaces the T-square and its position is maintained parallel to the top edge of the board by an arrangement of cords, pulleys, gears, or levers. Triangles and protractor are not needed since the left end of the straightedge may be pivoted and locked at any angle as measured by the protractor on the machine. This function is of special value when making pictorial drawings.

The drawing machine is a precision instrument; the pro-

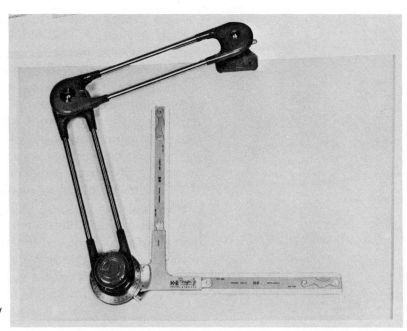

Figure 1-5 Drafting machine. (Courtesy of Keuffel & Esser Co.)

tractor degree dial at the lower left may be adjusted to a fraction of a degree. The vertical straightedge is maintained at 90° to the horizontal straightedge.

1-4
Pencils and
Line
Widths

Pencil leads are made of *graphite*. A special clay is added in different amounts to make 18 *grades of hardness* from 7B to 9H. The *soft* grades, 2B through 7B, are used for preliminary sketches since the lines produced are easier to erase. The *medium* grades, B through 3H, are used for general purpose work and lettering. The *hard* grades, 4H through 9H, are only used when extreme accuracy is required because the lines produced are apt to be too light. The final choice of pencil hardness also depends on the brand of pencil and the texture of the drawing paper.

The drafting pencil is usually sharpened to about 4 cm (1.5 in.) from the end, with about 10 mm (⅜ in.) of uncut lead exposed. The lead may be shaped to a sharp, conical point on a sandpaper block, or by a special hand sharpener for leads, and wiped clean. Special cutters or pencil sharpeners that remove only the wood to expose a given cylindrical length of lead are also obtainable.

Many draftsmen find it convenient and time saving to use a refillable pencil, or lead holder. Remember that only a *sharp* pencil is capable of producing clean-cut black lines (Fig. 1-6).

The actual width of lines is determined by the purpose, the size and style of the drawing, and the smallest size to which it is to be reproduced. Density of pencil lines depends on varying pressure applied by hand. Uniform pencil line widths necessitate constant and exact pointing of lead.

The *main visible* lines and *cutting-plane* lines should be the *thickest* or of greatest density as shown in Figures 1-6 and 1-11. *Hidden* lines should be of medium width. *Dimension* lines, *extension* lines, and *section* lines are usually drawn of *thin* density. *Construction* lines such as the orthographic projection lines drawn to obtain multiple views as shown in Figure 1-11, should be so *light* that they can barely be seen at arm's length. All final pencil lines should be clean, dense black, opaque, uniform, and spaced for legible reproduction.

**1-5
Erasers and
Erasing
Shields**

Erasers for removing either pencil or ink lines are available in many degrees of abrasion and hardness. *Artgum,* the softest grade, should only be used for general cleaning of large areas of a drawing or for removing unwanted pencil lines from an inked drawing. A soft pink eraser such as the Ruby eraser of Eberhard-Faber is preferred by many draftsmen for general work.

An *erasing shield* (Fig. 1-2b), is used to protect the lines near those being erased. The erasing shield has holes and slots of various sizes punched in a rectangle of thin sheet metal or plastic.

After prolonged erasing, a paper surface may become badly grooved; the surface can be improved by *burnishing* or rubbing with a hard, rounded, smooth metal object such as the metal cap of an automatic pencil.

**1-6
Inks and
Inking
Pens**

Some drawings are important enough to warrant the use of ink. In producing inked drawings, a pencil drawing may be traced through a tracing cloth or polyester film with inked lines drawn on the dull side.

With the advent of *microelectronics* and the necessity for

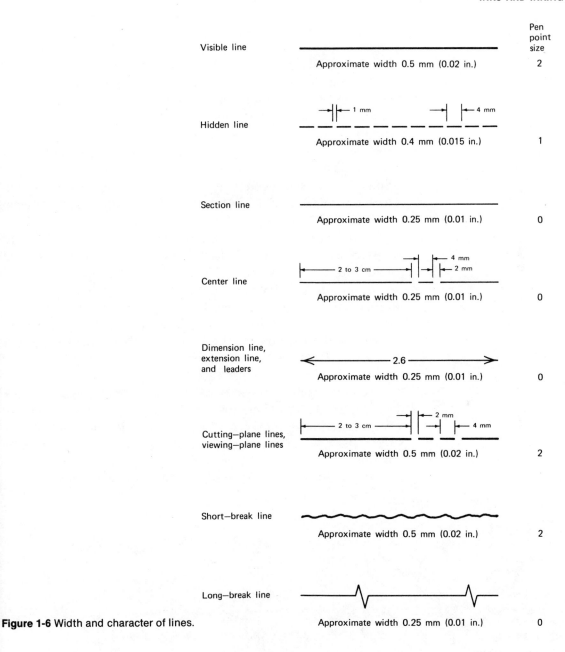

Figure 1-6 Width and character of lines.

photographic reduction of drawings, such drawings are being directly produced in ink.

For simple, straight line work in ink, the ordinary ruling or inking pen with adjustable parallel blades or *nibs* is used (Fig.

1-7). Line width is controlled by turning the knurled nut which regulates the distance between the tips.

Drawing inks are composed chiefly of finely divided carbon in colloidal suspension with gum. To load the pen with ink, first adjust the position of the blades until the tips are touching each other. The drawing ink is applied between the parallel blades using the *quill filler stem* (Fig. 1-7a) attached to the ink bottle stopper. A few lines are tried on scrap paper, and the knurled nut is adjusted for the desired line width. Excess or unused ink should never be allowed to dry in any inking pen. Clean the pen frequently with a stiff blotter or folded cloth placed between the nibs.

As with pencils, when drawing a line, the pen should be held so as to lean slightly in the direction of motion and in a vertical plane with respect to the drawing surface. Hold the pen so that the thumbscrew is faced *away* from the straightedge as shown in Figure 1-7b. The nibs should not be pressed too tightly to the straightedge since this action reduces the width of the line. Never tilt the pen toward the bottom of the straight edge of a triangle or T-square because the ink will run under the edge and mar the drawing.

Microelectronic drawings usually are drawn many times larger than full scale and must be photographically reduced for the production of *integrated circuits*. For this reason, microelectronic drawings are made in ink directly on *polyester* drafting film.

The technical fountain pen (Fig. 1-8b) was developed to ob-

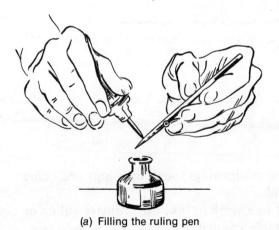

(a) Filling the ruling pen

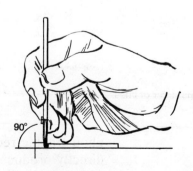

(b) Drawing position of the ruling pen

Figure 1-7 Inking techniques.

tain predetermined precise line width and corresponding uniform intensity at feather touch. The technical fountain pen draws curved lines of uniform density and width much faster and with greater ease than can be accomplished with the ruling pen.

(a) Cartridge type

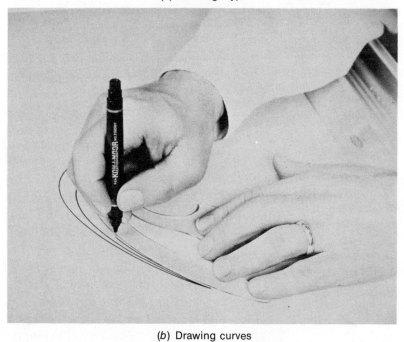

(b) Drawing curves

NIB

Figure 1-8 Technical fountain pen. (Courtesy of Koh-I-Noor Rapidograph, Inc.)

Point body

Capillary tube

Breather channel

Kolor—Koded Safety plug

Cleasing wire and stainless steel, Kolor—Koded weight

(c) Disassembled pen

Technical fountain pens are available in many popular, pre-determined tube point sizes, such as 0, 1, and 2 and methods of ink storage. Figure 1-8c shows a disassembled fountain pen. The cleansing wire floats freely within the point tube and is actuated by a precise weight so that it emerges below the tube orifice when the pen is held point down. The cleansing wire expels any drafting-surface accumulations. It also serves as an ink-feed regulator in the form of a valve when touching the drafting surface. A gentle, shaking motion will advance the ink prior to touchdown. The offset of the tip removes the danger of smearing ink lines.

Some technical fountain pens are filled by immersing the point and slowly turning a filler knob. The cartridge may be removed in the type shown in Figure 1-8a and filled with ink from a plastic squeeze bottle.

Ink on polyester film can be easily removed to make corrections or redrawings. Ink is retained on the film by adhesion to the surface *tooth* and not by penetration of the surface pores, as on a porous material such as paper. To erase inked details from drafting film, a special plastic eraser dampened with a liquid eraser has been developed. The liquid eraser softens the inked line while the friction of the solid eraser against the film surface removes the ink particles.

1-7 Instrument Kit

A typical set of traditional drawing instruments is included in the drafting kit shown in Figure 1-2c. A case or complete set of such instruments is a good investment. Individual or combination instruments also may be obtained as needed. By substituting needles for leads, for example, a compass may be converted into a divider.

High-quality instruments are usually made of corrosion-resistant nickel silver or stainless steel. A typical set contains a *compass, dividers, bow pen, bow pencil, bow dividers,* and two *ruling pens.* Instead of a separate compass for drawing small and large circles, some sets also include attachments that increase the versatility of a single compass. The distance between lead and point in one style is regulated by a thumb nut on the side of one compass leg; in another design, this distance is regulated by a thumb wheel centered between the legs.

Since mechanical drawings are drawn to scale with their dimensions proportional to the actual object, the draftsman also needs a *flat* or *triangular scale* as shown in the kit in Figure 1-2*d*. Scales are calibrated in either the English or metric system of measurement to as many as six proportions or scale reductions.

A *proportional divider* is a specialty instrument that may be added. Many draftsmen find that scale reductions or proportions are more quickly obtained with it than with a triangular scale. This instrument consists of two slotted metal bars with pointed ends coupled with a sliding pivot (Fig. 1-9). The pivot is adjusted so that the distance between the points at one end are in the desired proportion to the distance between the points at the other end.

1-8 Drawing Papers and Materials Electronic drawings are usually made on semitransparent materials so that the drawings are easily reproduced. *Natural bond, vellum, transparent cloth,* and *polyester film* are the most popular materials. The quality of these materials is judged on the basis of *weight, transparency, surface texture, erasing quality, stability, strength,* and *permanence.*

Standard available sheet sizes are listed in Appendix J. The *weight* of a paper is judged by the approximate weight in kilograms of one ream (500 sheets) of 43.2 × 55.9 cm sheets.

Natural bond tracing paper is an all-rag variety. It is the least expensive, but it lacks durability. Since the labor cost involved in producing a drawing is considerably more than the paper cost, it is unwise to use a poor-quality paper except for temporary work. Vellum is a higher quality rag paper having a greater transparency than natural bond paper. For greater permanence, coated transparent linen cloth is preferred.

The most durable and dimensionally stable material is polyester film. It has a matted surface to make the film suitable for both pencil and ink drawings. It can be obtained in thicknesses from 0.05–0.20 mm (0.002–0.008 in.)

Cross section papers, also known as *quadrille* or *coordinate* papers, are particularly useful for making preliminary sketches, schematic layouts, and wiring diagrams. Common graph papers having linear grid rulings of one and five lines per centimeter

Figure 1-9 Proportional divider. (Courtesy of Keuffel & Esser Co.)

and four to ten lines per inch in both directions are also available. Amplifier and control system characteristic curves are usually plotted on *semilog* papers in which one scale is logarithmic and the other is linear. *Log-log* papers are also available as well as special engineering papers such as *polar coordinate* papers, *Smith charts,* and *Nichols charts.*

Pictorial drawings are easily produced using a special cross-section paper that is imprinted with diagonal lines in two directions at about 30° to the horizontal lines.

1-9 Templates A drafting *template* is a thick, clear plastic sheet containing various precision cutouts of symbols and mechanical or geometrical shapes (Fig. 1-10). The use of templates saves drafting time and effort and achieves uniformity. Template thicknesses range from 0.5–2.0 mm (0.02–0.08 in.), with pencil point allowances on the cutout dimensions. Obviously the electronic symbol template is particularly useful to the electronic draftsman. Templates with various geometrical cutouts are also desirable. Other templates are available for electronic component outlines, circles, squares, rectangles, triangles, ellipses, computer gates, and lettering guides. Square, rectangle, and triangle outlines are helpful in making some block and system diagrams. Ellipse outlines are almost indispensible in developing pictorial drawings.

1-10 Drafting Aids Time and labor-saving devices also include adhesive, preprinted shapes and symbols. Those that are transparent are sometimes known as *appliqués.* A wide variety of electrical and electronic symbols, cross-section shading, and common abbreviation lettering can be obtained in stock sheets. The symbol or shape is cut from the sheet, after which the pressure-sensitive backing paper is removed before application to the drawing paper.

Opaque or silhouette adhesive shapes are often used to present printed circuit wiring. These drafting aids have the advantages of extreme uniformity, neatness, and high definition of intricate patterns in addition to saving drawing time.

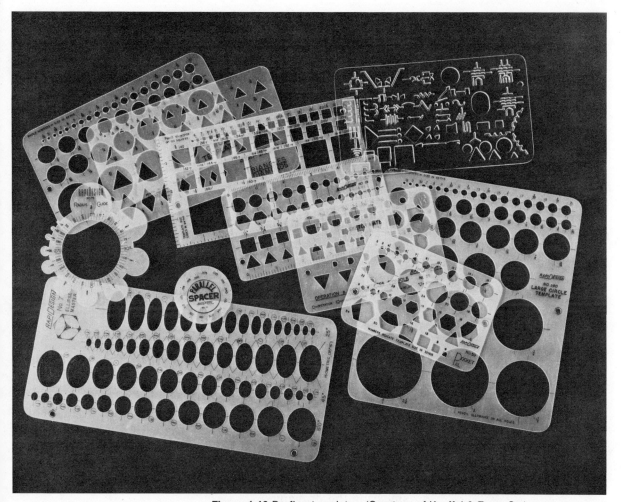

Figure 1-10 Drafting templates. (Courtesy of Keuffel & Esser Co.)

**1-11
Basics of
Orthographic
Drawing**

The T-square in conjunction with drafting triangles, are used to draw horizontal and vertical lines in a fundamental drafting procedure called *orthographic* or *multiview* projection as shown in Figure 1-11.

Any unsymmetrical three-dimensional object has six different *viewing planes*. The object of orthographic drawing is to show pertinent viewing planes by *unfolding* these planes by projection so that they will all appear on the same plane.

The view that shows the most detail is usually considered to be the *principal view*. The *natural front view* is not always the

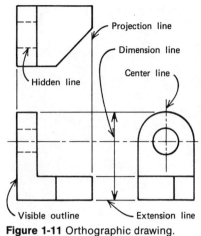

Figure 1-11 Orthographic drawing.

Projection line

Dimension line

Center line

Hidden line

Visible outline

Extension line

best one to use as the principal view. The principal view of the object is drawn as it ordinarily appears when viewed at a distance so that the *lines of sight* are parallel.

Three views, drawn to the same dimensional scale, are usually enough to fully describe the shape, size, and details of the object. The relative positions of the supplementary views should be positioned so as to appear *hinged* to the principal view through the *extension lines* as shown in Figure 1-11. To draw the side and top views, horizontal and vertical lines, respectively, are projected from the principal view. Obviously, the top view should be drawn above the principal view, and *side views* should be drawn to the left or right so that the drawing may be readily and accurately visualized by the observer.

Orthographic drawings require sufficient dimensioning to completely describe the object. Details such as radii, angles, holes, fillets, and the like require definition of sizes and location.

1-12
Work
Habits

Cleanliness is the most important habit in drafting work. Clean drawings only result from a conscious effort. Beginning with clean hands, all drafting equipment should be wiped frequently with a clean cloth. Graphite from careless pencil sharpening and the sliding of T squares and triangles across the pencil drawing may smudge the drawing. Avoid working with sleeves or arms resting on a penciled area. A dusting brush (Fig. 1-3e) may be used to remove eraser crumbs or bits of lead without smearing the drawing.

It is good practice to cover those parts of a drawing that are not actually being worked on at the time. The entire drawing should be covered with a clean paper or cloth at the end of a work period. Drawings should never be folded but should be stored flat in folders or in a drawing file cabinet.

It bears repeating to observe proper care of instruments:

1. Do not tighten inking pen beyond contact of nibs.
2. Do not force any threaded adjustments.
3. Relieve compass tension before storage.
4. Do not use dividers as picks or reamers.
5. Do not use a divider point as a compass center.
6. Do not permit ink to dry in inking pens.

7. Clean all instruments before returning to storage.

Summary **1.** The draftsman must be fully aware of the purpose and capabilities of conventional drawing equipment and materials.
2. Proficiency in the use of materials and drafting devices is essential.
3. The electronic draftsman should be familiar with templates and appliqués so as to prepare quality drawings in less time.
4. Good workmanship can be developed through practice, provided that the student pays attention to results.
5. An electronic drawing is essentially a means of communication and must clearly and completely conform to graphic principles and procedures.
6. Cleanliness is a particular virtue in drafting.

Problems **1-1** Construct three squares, 6 cm on a side. Evenly crosshatch each square at an angle of 45°, spaced at 2 mm, with each square containing lines of a different density.
1-2 Using drafting triangles and a T-square, accurately draw four triangles beginning with a 10 cm-long base and include the following angles:
(A) 90°, 75°, and 15°   (C) 120°, 45°, and 15°
(B) 45°, 60°, and 75°   (D) 105°, 30°, and 45°
1-3 Obtain a tubular capacitor from your instructor and make a two-view double scale drawing.
1-4 Obtain ½-, 1-, and 2-watt carbon resistors. Draw a two-view triple scale drawing of each.
1-5 Prepare a three-view orthographic drawing of a low-power transistor to a 10:1 scale.
1-6 Make a three-view orthographic drawing of a high-power transistor to double scale; include all dimensions.
1-7 Prepare a three-view orthographic drawing of a filament or other low power transformer to double scale; include all dimensions.
1-8 Make a three-view orthographic, full-scale drawing of an aluminum radio chassis 12 cm long, 5 cm wide, and 3 cm deep. Include all dimensions.
1-9 Prepare a full-scale sheet metal layout of the radio chassis of Problem 1-8. It should be laid out flat; the lines on

which it is to be folded should be shown with double dashes. Include four 6 mm-wide corner weld tabs and all dimensions.

1-10 One hundred 43.2 × 55.9 cm sheets of vellum are found to weigh 7.9 kg. What weight grade does this represent?

1-11 A printed circuit board is to be drawn to triple scale. Its actual dimensions are 6.3 × 8.4 cm. What will your actual drawing dimensions be?

1-12 What is the total weight of three reams of 86.4 × 112 cm vellum drawing paper? (See Problem 1-10)

Now return to the self-evaluation questions at the beginning of this chapter and see how well you can answer them. If you cannot answer certain questions, place a check next to them, and review the appropriate sections of the chapter to find the answers.

References **Charles J. Baer,** *Electrical and Electronics Drawing,* Third Edition, McGraw-Hill, New York, 1973, pp. 1–14.

Frederick E. Giesecke, Alva Mitchell, Henry Cecil Spencer, and **Ivan Leroy Hill.** *Technical Drawing,* Fifth Edition, Macmillan, New York, 1967, pp. 1–57.

Nicholas M. Raskhodoff, *Electronic Drafting and Design,* Second Edition, Prentice-Hall, Englewood Cliffs, N.J., 1966, pp. 1–58.

George Shiers, *Electronic Drafting,* Prentice-Hall, Englewood Cliffs, N.J., 1962, pp. 1–47.

Chapter 2 Lettering

Instructional Objectives

1. To understand the importance of lettering in electronics drafting.
2. To become familiar with styles of lettering.
3. To develop a proficiency in freehand lettering.
4. To learn how to plan word and letter spacing for best appearance.
5. To make you aware of the precautions necessary to produce superior lettering.
6. To provide the basic elements needed in the preparation of parts lists and tables.
7. To become familiar with the use of templates and mechanical lettering devices.

Self-Evaluation Questions

Test your prior knowledge of the information in this chapter by answering the following questions. Watch for the answers as you read the chapter. Your final evaluation of whether you understand the material is measured by your ability to answer these questions. When you have completed the chapter, return to this section and answer the questions again.

1. Why is good lettering important on an engineering drawing?
2. What *slope* is used for *inclined* lettering?
3. List three advantages of *uppercase* lettering.
4. How do the *curved* lines differ between *inclined* uppercase and *vertical* uppercase letters?
5. What is a *waistline?*
6. Name at least four precautions that must be observed for superior lettering *legibility*.
7. In lettering, what is an *ascender?*
8. What are the advantages in the use of *inclined* lettering?
9. Explain how *correct letter spacing* is obtained through an optical illusion.
10. Your *lettering template* may have thicker outer edges than the center portion. Explain the purpose of this design.

2-1
The Importance of Lettering

Lettering is an essential part of engineering drawings. Drawings contain information that the production and purchasing departments may need. Titles, part numbers, references, dimensions, component designations, wire identities, and component lists all require lettering. All lettering must be *simple* and *legible*. Poor lettering can ruin the appearance of a drawing. Illegible lettering can cause costly mistakes in the purchasing and manufacture of components and in the assembly of the device.

Although lettering machines, devices, templates, and special typewriters are available, the electronic draftsman should be able to produce neat and legible *freehand* lettering. Often, as much time is spent in lettering a drawing as in constructing the wiring diagram or assembly drawing. After you have become skillful in freehand lettering, you may choose to use a *lettering template*. The choice depends on your degree of proficiency and the amount of time expended. In either case, the electronic draftsman cannot expect to get along in professional work without being able to do neat, freehand professional lettering.

2-2
Letter Styles

Most electronic engineering department drawings require *single-stroke* Commercial Gothic-style lettering. The letters in this style may be *vertical* or *inclined*. These terms refer to the letter positions relative to the horizontal base *guideline* shown in Figure 2-6. In the *vertical* style, the centerline of the letter is perpendicular to the *baseline*. In the *inclined* style, the centerline of the letter is tilted to the right. Furthermore, *uppercase* or *capital* letters, and *lowercase* or "small" letters may be used. Usually uppercase lettering is preferred by most engineering departments. These standardized alphabets are presented in Figures 2-1 and 2-2.

2-2.1
Vertical Gothic Lettering

To show the comparative width of letters and position of lines composing each letter, the letters and numerals in Figure 2-1 are placed in blocks that are six small squares high. Note that the width varies according to the letter. The crossbar of the H and the middle bar of the E and F are drawn slightly above center. *Serifs,* the stroke crossing or projecting from the main stroke of a letter, are never used in vertical or inclined Gothic

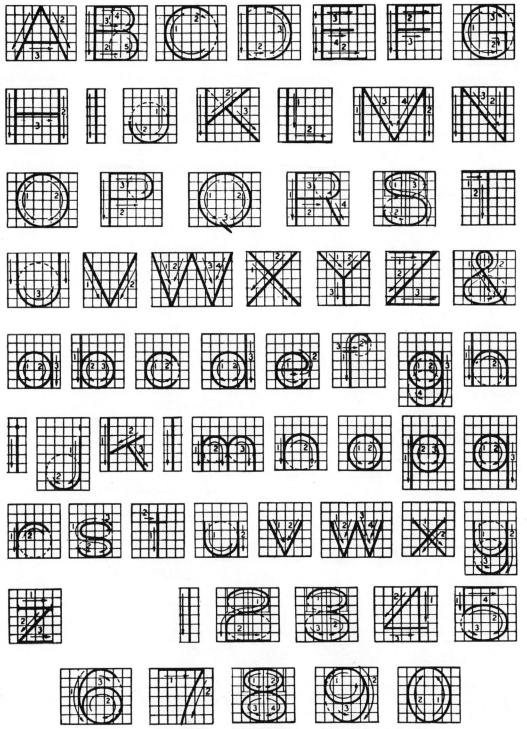

Figure 2-1 Vertical Gothic lettering style.

Figure 2-2 Inclined Gothic lettering style.

lettering. Curved lines are either circular or elliptical. Note that the uppercase letter O is a *circle* while the *zero* is an *ellipse*. Numerals are usually made narrower than letters. If the student is right-handed, the vertical lines are usually drawn from the top to the bottom and the horizontal lines are drawn from the left to the right.

2-2.2
**Inclined
Gothic
Lettering**

Inclined lettering (Fig. 2-2) is generally drawn at a slope of about 67.5°. The general proportion of characters for inclined lettering are the same as for vertical lettering.

2-2.3
**Microfont
Lettering**

A new type of lettering known as the *Microfont open style* is being adopted by many draftsmen because it reproduces very well on microfilm. In this style (Fig. 2-3), the design of some of the letters and numbers have been changed from the pure Gothic type for clearer reproduction and to prevent confusion; for instance, the Gothic number 4 is sometimes confused with the number 9.

ABCDEFGHIJKLMNO
PQRSTUVWXYZ.,:;
1234567890=÷+−±
@&⁂?#×"%'()[]° !
¢$/_∠∞∆≈~∅⊥<>μα
√σδΣΥΠβθωΩ∴‖

4 MM

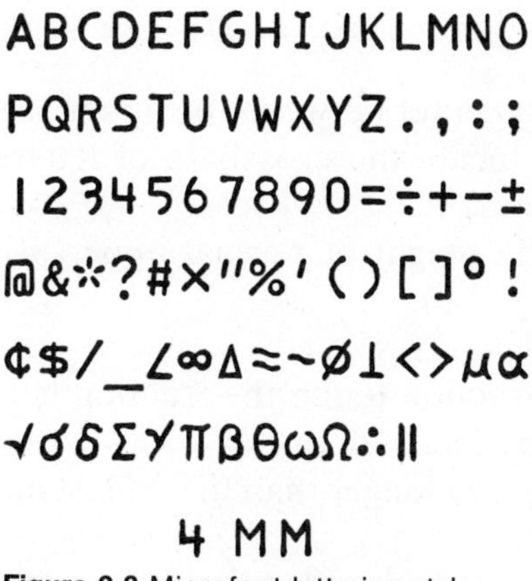

Figure 2-3 Microfont lettering style.

The *same* lettering style should be used consistently on any given drawing. All letters and numerals should consist of dense, black, medium-weight strokes. A minimum height of 0.4 cm ($5/32$ in.) is specified for all military drawings (MIL-STD-1). All horizontal lettering should be readable from the bottom of the sheet. Vertical and angular lettering should be readable from the bottom or right-hand side of the drawing. Correct spelling is always important; a technical dictionary should be used when in doubt.

2-3
**Guidelines and
Guideline
Devices**

Very light, thin guidelines should always be drawn to locate the top and bottom of all freehand uppercase lettering. Guidelines are conventionally drawn with the aid of a scale and T-square. Clear plastic templates provided with various widths of cut out slots may be used to quickly draw the desired guidelines (Fig. 2-4). The engraved lines above the slots are helpful in spacing lines of lettering.

Another lettering guide device consists of a two-piece template having a rotating disk. The disk contains holes for pencil points. As the disk is turned, the guideline spacing is changed.

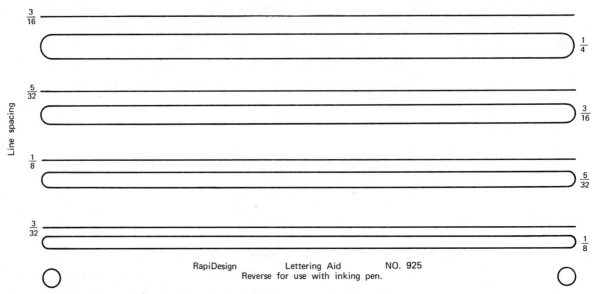

Line spacing

$\frac{3}{16}$

$\frac{1}{4}$

$\frac{5}{32}$

$\frac{3}{16}$

$\frac{1}{8}$

$\frac{5}{32}$

$\frac{3}{32}$

$\frac{1}{8}$

RapiDesign   Lettering Aid   NO. 925
Reverse for use with inking pen.

Figure 2-4 Guideline template. (Courtesy Bruning Div. AM.)

1. A guideline drawn in the exact center between the upper and lower guidelines is useful to locate the crossbars of letters and the division bar of fractions.
2. Fractions are drawn twice the height of normal uppercase letters or whole numbers (Fig. 2-5).
3. Note that the guidelines are *equally* spaced.
4. Never let the numerals of a fraction touch the fraction bar.
5. The fraction bar should *not* be inclined.
6. Draw the fraction bar only *slightly* longer than the widest numeral.
7. The denominator should be centered under the numerator.

For lowercase letters, it is advisable to draw additional guidelines known as *waistlines* and *droplines*. Figure 2-6 shows that the waistline determines the top of the lowercase body, while the dropline determines the length of the *descenders*. The descender length should equal the *ascender* length. The *cap* and *baselines* are the same guidelines as used for uppercase letters.

**2-4
Spacing of
Letters**

The three possible methods of horizontal letter spacing are shown in Figure 2-7. It should be obvious that the bottom spacing has the most pleasing appearance. The distance

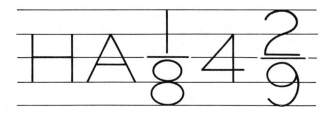

Figure 2-5 Fraction guidelines.

Cap line
Waistline
Baseline
Dropline

Figure 2-6 Lowercase letter guidelines.

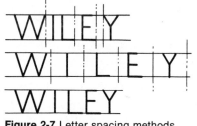

Figure 2-7 Letter spacing methods.

between each letter of a word is kept constant in the top arrangement. The same amount of space is given to each letter in the center arrangement. For a pleasing appearance, the volume of open space should be the same as in the lower arrangement. All letters should appear to have equal spaces between them. Note that when the sides of adjacent letters are vertical lines, the spacing between these letters is the greatest. Make the spaces between words approximately equal to the width of an uppercase O.

In lettering such as *titles* or where more than one line of lettering is to be used, the title word (or words) must be *centered* within the *title space*. The clear distance between lines is made from one-half to one and one-half times the height of the uppercase letters (Fig. 2-8). Make trial layouts of the lettering on a separate piece of paper until the lettering is symmetrical about a vertical centerline. Since the most pleasing spacing of letters is determined by the shape of adjacent letters, symmetry is usually not achieved by counting the letters and spaces between words.

**2-5
Lettering
Templates**
A lettering template is a form of metal or plastic stencil containing narrow slots and holes corresponding to the outlines of all letters and numerals. Most of the letters completely perforate the stencil; the pencil or lettering pen point is moved along the inside edges of the perforation. The pencil must always be held at right angles to the template in all directions. Lettering

guidelines may not be necessary if the template is moved horizonally while in contact with a T-square edge to produce succeeding letters. To be able to draw such letters as A, B, K, P, and R, it is usually necessary to draw the letter using two separate perforated shapes. Obviously, a separate stencil is needed for each height and style of lettering.

Special precautions must be taken when using a lettering pen. It is best to use lettering templates that are molded with raised edges to prevent the letter perforations from touching the drawing paper. Use of such templates helps prevent ink smudges.

Inside the lettering pen is a cleaning pin used to keep the small tube point open (Fig. 1-8). Cleaning pins are easily broken if the pen is not properly cleaned. To clean a tube-type lettering pen:

1. Draw it across a blotter until all ink has been absorbed.
2. Insert the cleaning pin.
3. Remove the cleaning pin and wipe it with a cloth.
4. Repeat the last two procedures until the pin is completely clean.

Use of a lettering template helps achieve maximum clarity and uniformity. It prevents the draftsman from adding individual characteristics or flourishes. It is especially important for maintaining *uniform* lettering when a drawing is being lettered by more than one draftsman.

2-6 Parts Lists and Tables The title of a drawing and related information are lettered in *title boxes* or strips located in the lower right-hand corner of the drawing sheet. Usually these boxes and a border are *printed* with the company name by the supplier of the drafting paper. Figure 2-8 shows several variations of *title boxes* and also a *parts list* or *bill of material*. In industry, companies develop title boxes and parts lists to suit their own needs. The following items are usually contained in a title box:

1. Name of company.
2. Title or exact description of object.
3. Name and part number of the assembly into which the part is designed to fit.
4. Who made the drawing.

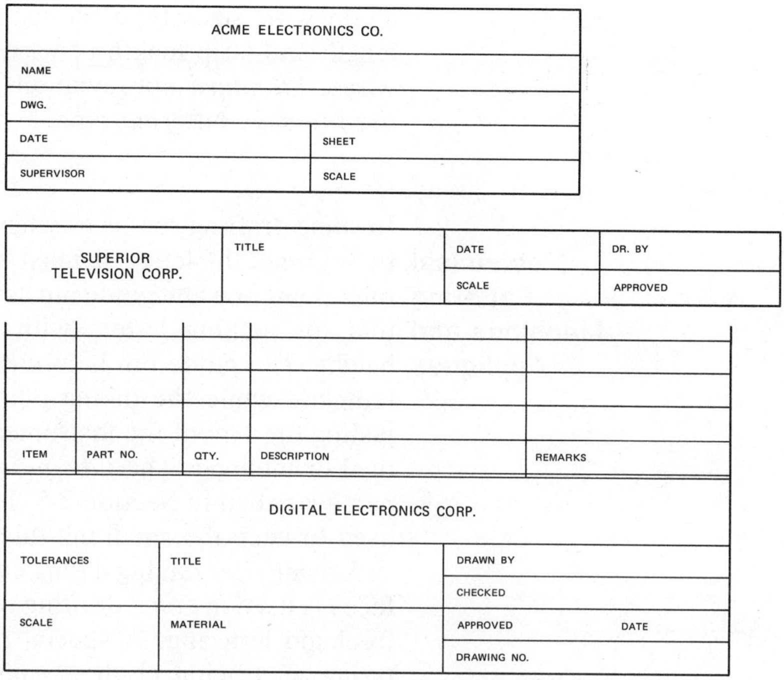

ACME ELECTRONICS CO.	
NAME	
DWG.	
DATE	SHEET
SUPERVISOR	SCALE

SUPERIOR TELEVISION CORP.	TITLE	DATE	DR. BY
		SCALE	APPROVED

ITEM	PART NO.	QTY.	DESCRIPTION		REMARKS
ITEM	PART NO.	QTY.	DESCRIPTION		REMARKS

DIGITAL ELECTRONICS CORP.			
TOLERANCES	TITLE	DRAWN BY	
		CHECKED	
SCALE	MATERIAL	APPROVED	DATE
		DRAWING NO.	

Figure 2-8 Title boxes and parts list.

5. When was the drawing prepared.
6. Approval by supervisor.
7. Scale.
8. Revision record.

Lettering in a title box must be balanced or arranged symmetrically about an imaginary vertical centerline. Uppercase letters are almost always used. The more important words are made with larger or heavier lettering than such data as the scale, date, and the like.

Parts lists may either be drawn immediately above the title box as in Figure 2-8 or in the upper right-hand corner of the drawing sheet. The number of columns is determined by the type of drawing. If the drawing concerns the manufacture of a component or electrical part, the list must contain the material from which the component was fabricated, such as type of material, thickness, and finish. On assembly drawings, the number of identical components, the part number of the component and the description and specifications of the component are necessary information. Wiring harness parts lists should include wire size, type of insulation, color of insulation, cut length and strip length of wires, part number, and location of wires. *Standard* abbreviations given in Appendix A may be used to save lettering time.

2-7 Mechanical Lettering Machines and Appliques In some drafting rooms or when drawings are to be reproduced by printing, the lettering must also be done with pen and ink. A tube-point pen and guide pin assembly (Fig. 2-9) is a convenient tool for making letters with strokes of uniform width and height. The guide pin is moved to follow grooved letters in a template while the inking point moves on the paper. By adjusting the arm of the instrument, the letters may be made vertical or inclined. The *tube-point pen* is similar to the fountain pen described in Section 2-5. Inside the tube-point pen is a pin used to keep the small ink tube open.

A *typewriter* having a long carriage and interchangeable type faces is used in some drafting rooms to reduce the labor cost of freehand lettering. A special ribbon is used to give clear lettering on tracing cloth. A miniature typewriter, mounted in place of the horizontal guide of a drafting machine, is also available.

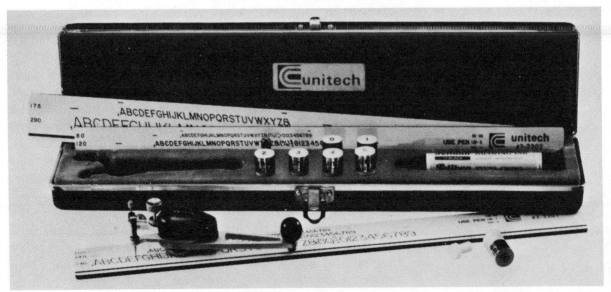

Figure 2-9 Tube-point lettering instrument and templates. (Courtesy Bruning Div. AM.)

Another method of reducing the labor cost is machine printing of commonly used information on transparent paper having an adhesive backing. This method, called *appliqué,* is also useful for the rapid application of frequently used electronic symbols.

Summary
1. Good lettering breeds confidence in a drawing.
2. Good legibility prevents costly mistakes.
3. The vertical uppercase Gothic style of lettering is preferred, but consistency of style is most important.
4. Skill in freehand lettering is developed through practice.
5. Very light guidelines are always used in freehand lettering.
6. The spacing of letters, words, and titles is important for reasons of legibility.
7. The use of lettering templates, mechanical lettering devices, and appliqués may be of value as labor-saving devices or when freehand lettering skills have not been properly developed.
8. Parts lists and tables must be legible and contain all component specifications to prevent mistakes in purchasing and manufacture.

Problems
2-1 Print your name, social security number, school name, city, state, and date in centered tabular form using 7 mm inclined Gothic uppercase letters.

2-2 Print all the letters of the alphabet using 10-mm vertical Gothic uppercase letters in two lines, and all numerals in a single line in proper order. Pay strict attention to the proper spacing between adjacent letters and numerals as shown on the bottom line of Figure 2-7.

2-3 Lay out and letter this course number, course name, and your instructor's name in a centered, three-line tabular form. The lettering should be 0.5-cm vertical upper- and lowercase Gothic, allowing a clear distance of 4 mm.

2-4 Design, draw, and letter a title box containing the following information: Phoenix Radio Corp., cold-rolled steel cadmium-plated chassis, for a model F-62 radio, drawn by yourself at half-scale on February 16, 1975, and checked by your instructor. Use 6-mm uppercase Gothic for the company name and 4-mm uppercase for the balance of the lettering. Standard abbreviations may be used.

2-5 Using the data from the last problem, prepare the title box using 0.7-cm Microfont lettering for the company name and 0.4-cm Microfont lettering for the balance of the lettering.

2-6 Plan a parts list box in which there are nine lines including the title line. All lettering is to be 4 mm high. The clear distance should be 3 mm. The longest line should read "3 C-342 CAPACITOR 1 μF 25 WVDC." Determine the overall size of the box.

2-7 Obtain a small printed circuit board containing mounted components from your instructor. Prepare a complete parts list using 5-mm vertical uppercase Gothic lettering.

Now return to the self-evaluation questions at the beginning of this chapter and see how well you can answer them. If you cannot answer certain questions, place a check next to them, and review appropriate sections of the chapter to find the answers.

References **Charles J. Baer,** *Electrical and Electronics Drawing,* Third Edition, McGraw-Hill, New York, 1973, pp. 6–13.

Thomas E. French and **Charles J. Vierck,** *Engineering Drawing,* Tenth Edition, McGraw-Hill, New York, 1957, pp. 83–93.

Frederick E. Giesecke, Alva Mitchell, Henry Cecil Spencer, and **Ivan Leroy Hill,** *Technical Drawing,* Fifth Edition, Macmillan, New York, 1967, pp. 62–85.

George Shiers, *Electronic Drafting,* Prentice-Hall, Englewood Cliffs, N.J., 1962, pp. 169–198.

Chapter 3

**Block
and
Logic
Diagrams**

Instructional 1. To understand purposes of *block, flow,* and *logic* diagrams.
Objectives 2. To become familiar with symbols used in preparation of block and
logic diagrams.
3. To learn how to plan an arrangement of block symbols to produce
an intelligible block or flow diagram.
4. To recommend proper drafting procedures for preparation of easily
understood block diagrams.
5. To provide the basic elements of logic symbols, diagrams, and
Boolean expressions.
6. To understand operation of a logic diagram through use of a *truth
table.*

Self-Evaluation Test your prior knowledge of the information in this chapter by
Questions answering the following questions. Watch for the answers as
you read the chapter. Your final evaluation of whether you understand the material is measured by your ability to answer
these questions. When you have completed the chapter, return
to this section and answer the questions again.

1. Why use block diagrams?
2. How do computer program flow diagrams generally differ from
electronic circuit block diagrams?
3. Name at least three precautions in drawing block diagrams.
4. Where are auxiliary stages, such as oscillators, usually placed in
space relationship to the main stages?
5. How is a flow diagram drawn to permit repetition of a portion of an
instruction sequence?
6. What is the advantage of using distinctive shapes, rather than
square blocks, in a logic diagram?
7. What is the function of an AND logic gate?
8. How is the output of a logic diagram determined?

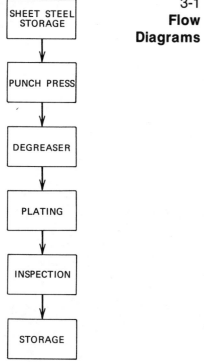

Figure 3-1 Chassis fabrication flow-chart.

3-1
Flow
Diagrams

A *flow* diagram or chart shows the sequence of operations or *stages* for a computer program or an industrial process. The name may have been derived from various plans for the flow of water. The name of each operation or stage in a flow diagram is contained in a box, or block, to simulate a process or area of operation. The blocks are usually rectangular and are connected with single lines to indicate the *sequence of operations*. The sequence of operations in a *flowchart* usually progresses from the *top* of the sheet to the *bottom*. Figure 3-1 shows a flowchart for the steps in the manufacture of an electronic *chassis*. It begins with the raw sheet steel and ends with the delivery of the plated chassis to a storage area. The plating department block and almost every other block may be further broken down into a *departmental* flowchart.

The blocks should all be drawn to the same dimensions. The block containing the most lettering usually determines the size of all the blocks in a diagram. All lettering should be of the same style and height. The arrangement of blocks should be centered along an imaginary line. The *flow lines* are drawn with arrows to indicate the direction of flow between operation blocks. The arrows are *usually* drawn at the *input* side of the blocks but may also be *centered* on a flow line. The flow lines should not be staggered but must follow each other vertically unless the chart is drawn to illustrate multiple flow paths. When more than one flow line emerges from the side of a block, the lines should be spaced symmetrically.

Figures 3-2 and 3-3 illustrate how flowcharts are used in writing a computer program. When it is necessary to solve the same problem many times using different sets of data, the computer is especially useful, since the operations are performed at extremely high speeds. The complete computer diagram in Figure 3-2 involves *double* flow so as to *branch, skip,* or *jump* instructions. The computer is then able to repeat the same sequence of instructions without being stopped and restarted. Note that the flow lines are drawn either vertical or horizontal; *diagonal* flow lines are *rarely* used. The blocks and the white areas are arranged to achieve a symmetrical pleasing appearance.

The logic and arithmetic stages of a computer are detailed in the flow diagram of Figure 3-3 to explain the first step in an evaluation program of an algebraic expression or formula. An

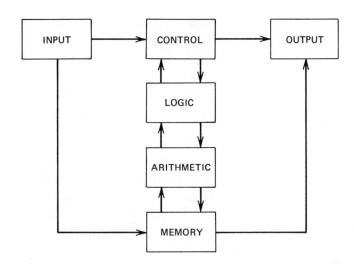

Figure 3-2 Computer flow diagram.

algebraic formula is an expression of how several *variables* relate to each other. The flowchart in Figure 3-3 shows the sequence in which the mathematical operations are performed for solution of a quadratic equation. The input stage substitutes numerical data for the alphabetical characters, while the arithmetic stages perform the indicated addition and multiplication.

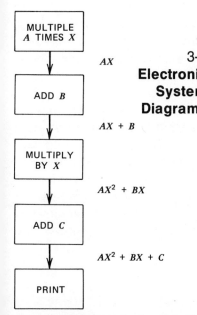

Figure 3-3 Flowchart for the evaluation of an algebraic expression.

3-2
Electronic
System
Diagrams

The distinction between flow diagrams and block diagrams is vague since common usage dictates the name given the diagram by a particular industry. In electronics, a left-to-right convention exists, and these system drawings are normally called *block diagrams*.

The simplest electrical block diagram or single-line diagram is intended to describe the basic *functions* of a circuit or system. It usually does *not* include the detailed information and individual parts identification of the *schematic diagram*. The emphasis is on the function of each stage in a system rather than on the composition of the stage.

Block diagrams may be the first step in an engineering design. With the present availability of integrated circuits or electronic modules, wiring diagrams are becoming a combination of block and schematic diagrams. Block diagrams are often used in electronic catalogs and sales literature, since the general functions of the device are better understood by persons not trained in electronics. The electrical wiring layout for a building uses a combination of the block and schematic diagram, but the devices are connected with single lines that may represent two to four conductors. (See Chapter 11.)

Figure 3-4 illustrates a very simple single-line block diagram of a tuned radio frequency (TRF) receiver. By convention, the input or signal is drawn at the left side of the diagram. The signal progresses through amplification and detection horizontally to the right or output side of the diagram.

A given block may represent a complete and removable chassis or system, such as an amplifier or a TV camera. The block diagram may become more detailed in that each symbol represents a *stage* as in Figure 3-4. A stage usually consists of an *active* component, such as a vacuum tube or transistor, together with its associated *passive* components, such as resistors and capacitors.

The drawing of a block diagram by stages is usually followed by a *schematic diagram* of each stage in which all component symbols are included. In the last few years, the schematic diagram has partially returned to the block diagram when *integrated circuits* (ICs) are included in the circuit, as shown in Figure 5-8. The IC may itself contain dozens of transistors, diodes, and resistors. In this instance, the IC symbol may be a block or triangle connected to external passive components, such as capacitors, inductors, and resistors.

Figure 3-5 shows a modern superheterodyne radio receiver and indicates *feedback* flow lines in addition to the conventional left-to-right signal flow. Note that the schematic symbol for the antenna and the speaker is added to this block diagram. Also note that the radio frequency (RF) is combined with the output of the oscillator, after which the intermediate frequency (IF) signal is amplified. The detector stage both rectifies and removes the IF portion of the signal and passes the audio frequency (AF) or intelligence component to be amplified. The

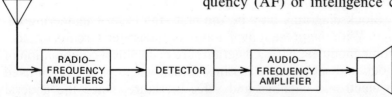

Figure 3-4 Tuned radio frequency (TRF) radio block diagram.

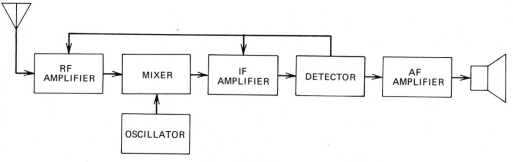

Figure 3-5 Superheterodyne radio block diagram.

amplified signal is then converted into sound by the speaker. A small portion of the AF signal is filtered to obtain a varying direct-current (dc) potential which is fed back to the RF and IF amplifiers in order to automatically control the gain (AGC) or amplification equalized for nearby strong or distant weak stations. (See Figure 3-5.)

A recent electronic development is the *electronic watch* having *digital readout* of time. The electronic watch circuit in Figure 3-7 is accurately controlled by the precision vibration mode of a *quartz crystal*. The controlled oscillator, frequency divider, wave shaper, and decoder blocks or stages are all combined in a single IC module.

The *overall layout* of any block diagram should be drawn to obtain pictorial balance or symmetry (See Figs. 3-5, 3-6, and 3-7). The *spacing* of the blocks is arranged with about the *same* amount of open space *between* adjacent blocks. The dimensions of the blocks is again determined by the maximum legend to be contained within the block.

The individual lines connecting block diagrams may represent single or multiple conductors. Diagonal lines are seldom used since they detract from the appearance of the diagram. Where many flow lines emerge from or enter a single block, grouping of lines and arrowheads should be avoided for the sake of legibility and appearance (Fig. 5-8d). Parallel flow lines should be spaced as far apart as possible. Where parallel flow lines cannot be avoided, arrowheads should be staggered or arranged in a geometrical pattern. Input lines may be drawn to enter more than one side of the block if appearance and legibility of the block diagram are improved.

BLOCK AND LOGIC DIAGRAMS

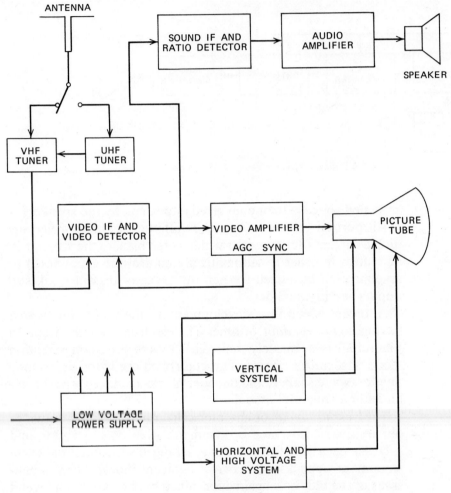

Figure 3-6 Block diagram of a VHF-UHF TV receiver.

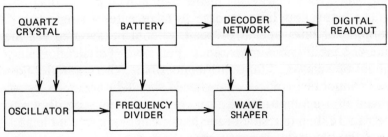

Figure 3-7 Electronic watch block diagram.

**3-3
Digital
Logic
Diagrams**
The digital logic designer builds a computer or control circuit by starting with a few types of simple, basic circuits, called *logic blocks*. Many blocks are interconnected to perform the various computer functions.

Although the numerical input data to a computer are decimal numbers, the logic operations within the computer are performed in the *binary number system*. The binary number system uses only two digits, 0 and 1. An electrical circuit is basically binary in nature in that the circuit is either turned on or off. The desire for reliability led designers to use switches, relays, and finally transistors, since these devices cause the circuit to operate essentially in one of two states, fully conducting (ON) or nonconducting (OFF).

Logic blocks have one or more input lines but only one output line. Some basic logic block symbols are given in Figure 3-8. A logic block is defined by giving the output condition for every possible combination of input conditions. Such input–output characteristics are tabulated in a *truth table*. Note that the name of a logic block refers to the input–output characteristics. For example, the AND logic gate does not have a 1 output until *both* A *and* B input lines are at a 1 level (Fig. 3-8*a*). Similarly, the OR logic gate (Fig. 3-8*b*) has a 1 output when a 1 level signal or pulse is applied to *either* the A *or* B inputs.

The basic shape of the *inverter block* in Figure 3-8*c* is a triangle. The triangle symbol by itself indicates that it is an amplifier; however, the tiny circle (or bubble) at the output symbolizes that the amplified output is inverted or reversed in *polarity*. For this reason, sometimes an inverter logic gate is called a NOT gate. One side of the triangle is always drawn vertically and is considered the input side of the block.

Note that in the NAND logic gate symbol (Fig. 3-8 *e*), the output of an AND gate is inverted by the small bubble at the output. The *acronym* NAND means *not* AND.

The letters shown at the outputs of the gates in Figure 3-8 are statements in *Boolean algebra*. Operations in Boolean algebra are somewhat different from the operations in ordinary algebra. The *variables* are denoted by letters but take only one of two *logic levels* or values, either 0 or 1. An AND operation is indicated by placing the two letters next to each other; this positioning of the letters does *not* indicate multiplication as in ordinary algebra. The OR operation differs from the AND

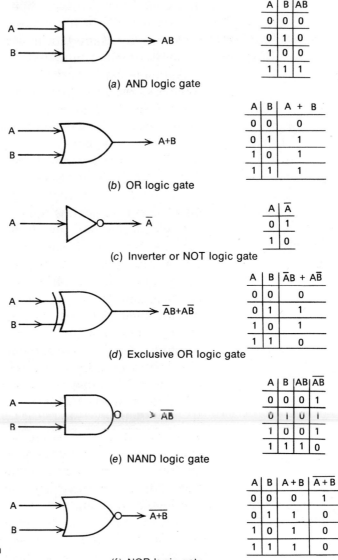

(a) AND logic gate

A	B	AB
0	0	0
0	1	0
1	0	0
1	1	1

(b) OR logic gate

A	B	A + B
0	0	0
0	1	1
1	0	1
1	1	1

(c) Inverter or NOT logic gate

A	$\overline{A}$
0	1
1	0

(d) Exclusive OR logic gate

A	B	$\overline{A}B + A\overline{B}$
0	0	0
0	1	1
1	0	1
1	1	0

(e) NAND logic gate

A	B	AB	$\overline{AB}$
0	0	0	1
0	1	0	1
1	0	0	1
1	1	1	0

(f) NOR logic gate

A	B	A+B	$\overline{A+B}$
0	0	0	1
0	1	1	0
1	0	1	0
1	1	1	0

Figure 3-8 Logic symbols and truth tables.

operation by the separation of the letters with a plus symbol. The plus symbol does not mean addition as in ordinary algebra.

A bar placed over a letter or combination of letters indicates that its value is inverted or has the opposite logic level of the letter without a bar. For example, refer to the right-hand column $\overline{AB}$ of the truth table in Figure 3-8e in which all outputs are of opposite value to the outputs in the third column (AB).

The conventions observed in the foregoing description are used in *positive logic* diagrams; in *negative logic,* these conventions are reversed. For example, when a small circle is drawn on the input side of a gate, the polarity of the input signal is reversed or inverted before entering the basic logic gate.

Other logic gates have *flip-flop* circuitry for counting circuits, shift registers, and memories; these gates are symbolized by properly identified rectangular blocks. Within the last few years, solid-state component manufacturers have combined many logic gates into single modules called *integrated circuits* (ICs). Such ICs are drawn simply as rectangular blocks identified by the manufacturer's number with an appropriate acronym to indicate function.

Summary 1. Block and flow diagrams may be the first step in a process or product design.
2. Sales and service brochures often use block and flow diagrams because of their simplicity.
3. Rectangular boxes, or blocks, representing electronic stages or steps in a process, are connected by flow lines in the order in which the operation occurs.
4. By convention, the initial block is drawn at the top or upper left corner of a diagram.
5. All blocks must be identified with a legible and symmetrically placed legend.
6. Blocks and flow lines must be equally spaced to present a pleasing drawing.
7. Computer circuit diagrams use slightly more distinctive blocks to represent logic gates or electronic circuits.
8. An understanding of computer data flow is simplified through the use of truth tables and Boolean algebra.

Problems 3-1 Draw a flowchart to program the algebraic expression $3(X + 5)$.

3-2 Prepare a flowchart to program the algebraic expression $2X^3 + X^2 + 4X + 3$.

3-3 Construct a block diagram showing the flow of power

from a three wire input power line to the service center and distribution to a furnace, water heater, clothes dryer, and lighting load.

3-4 Construct a flow diagram of the departments involved in putting a new electronic device into production. Include blocks for purchasing, drafting, engineering design, inspection, sales, quality control, assembly, and shipping.

3-5 Prepare a block diagram of an automobile electrical system. Include a block for the alternator, battery, lighting system, ignition system, and starter.

3-6 Draw a block diagram of a tape recorder. The stages should progress in order from the microphone, through an amplifier, to the recording head. Include a record-playback control block (that will substitute a playback head for the microphone) and a speaker for the recording head.

3-7 Prepare a block diagram of an AM–FM combination radio using a single AF amplifier and speaker. Refer to the AM superheterodyne radio receiver shown in Figure 3-5.

3-8 Feed the output of an OR logic gate to an inverter logic gate. Prepare a truth table for the gate combination. How does the output of the combination compare with other logic gates?

3-9 Draw an AND logic gate; connect its output to one of the inputs of an OR gate. Label the other OR gate input as C. Prepare a truth table for all possible combinations of data applied to the three inputs.

3-10 Reverse the logic circuit of Problem 3-9, connecting the OR logic gate output to one of the two AND gate inputs. Determine the output of this combination by preparing a truth table.

Now return to the self-evaluation questions at the beginning of this chapter and see how well you can answer them. If you cannot answer certain questions, place a check next to them, and review appropriate sections of the chapter to find the answers.

References **Charles J. Baer,** *Electrical and Electronics Drawing,* Third Edition, McGraw-Hill, New York, 1973, pp. 108–122.

Thomas C. Bartee *Digital Computer Fundamentals,* Third Edition, McGraw-Hill, New York, 1972, pp. 6–25, 79–90.

Gerhard E. Hoernes and **Melvin F. Heilweil,** *Introduction to Boolean Algebra and Logic Design,* McGraw-Hill, New York, 1964.

Chapter 4 Electronic Component Symbols

Instructional Objectives
1. To understand the need for electronic symbols and schematic diagrams.
2. To become familiar with basic functions of commonly used electronic components.
3. To make you aware of symbolic similarities.
4. To relate component symbol shape to component function.
5. To develop a proficiency in drawing electronic symbols in acceptable standard form.
6. To become familiar with the *resistor color code*.
7. To learn how to use *symbol templates, grid paper*, and *appliqués* in drawing schematic diagrams.

Self-Evaluation Questions Test your prior knowledge of the information in this chapter by answering the following questions. Watch for the answers as you read the chapter. Your final evaluation of whether you understand the material is measured by your ability to answer these questions. When you have completed the chapter, return to this section and answer the questions again.

1. What is the purpose of an electronic symbol?
2. How does an electronic component relate to its symbol?
3. Explain the basic similarities of symbols for *rheostats, potentiometers*, and *resistors*.
4. What specifications does the resistor color code furnish?
5. How does the physical size of a *transistor* symbol compare with the size of an *electron tube* symbol?
6. What is the difference between a *unijunction* transistor and a *field-effect transistor* symbol?
7. How does an SCR symbol differ from the symbol for a semiconductor diode?
8. What does the following resistor color code represent: red, blue, yellow, and silver?
9. What is the difference between an *npn* and a *pnp* transistor symbol?

10. What does the *acronymn* DPST represent?

11. How can the *emitter* element of a transistor symbol be identified?

12. How does a *radio-frequency transformer* symbol differ from an audio-frequency transformer symbol?

13. What specifications must be contained in the parts list for a *capacitor?*

4-1
The Development of Electronic Symbols

A symbol is a graphical pattern or pictorial device used to represent something else; in electronics, symbols are a technical shorthand. It is often quite difficult to determine *stage function* in a hand-wired chassis or printed circuit board containing the actual components. The function of individual stages is quickly determined by a competent electronic technician from a wiring diagram containing symbols to represent components.

Originally, symbols were drawn to look something like the components they represented. However, many components are now packaged or modularized, thus preventing easy identification. Symbols have been simplified and mainly relate to the basic function of the device rather than its appearance. It has therefore become more important that the technician understand the function and capabilities of electronic components when trying to read a wiring or *schematic* diagram.

Representatives of industry and government have almost agreed on the standardization of symbols and specifications in electronics. Refer to Appendix F for the titles and numbers of government and industrial publications concerned with electronic standards.

4-2
Inductors

An *inductor* serves to oppose a change in current flow in an electrical circuit. A practical inductor consists of a coil of several turns of insulated wire on either an air, *ferrite,* or *iron core*. The basic inductor symbol therefore consists of a *helical,* or corkscrew, line as in Figure 4-1a. *Chokes, transformers,* and *relays* are applications of inductors and are similar in the shape of their symbols.

Figure 4-1b illustrates a radio-frequency (RF) choke. The dashed line indicates that the inductor has a *ferrite* core; an arrow at the end of the dashed line indicates that the position of the core is adjustable. In Figure 4-1c the solid lines mean that the wire is wound on a fixed *laminated iron* core; this symbol is

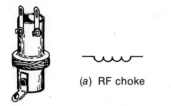

(a) RF choke

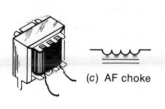

(b) Adjustable RF choke

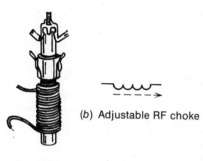

(c) AF choke

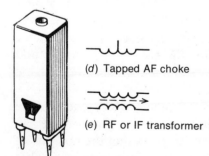

(d) Tapped AF choke

(e) RF or IF transformer

Figure 4-1 Inductor symbols.

used when the choke is designed for use in a low- or audio-frequency (AF) circuit. Figure 4-1d illustrates a *tapped* radio-frequency choke with an air core. Typical pictorial illustrations of the components are shown adjacent to their symbols.

The parts list specifications for a choke must include its *inductance* value and its maximum current-carrying capacity. Inductance is measured in *henrys* (H), *millihenrys* (mH), or *microhenrys* (μH) and is directly related to the square of the number of turns of wire in the winding.

Transformers may consist of a tapped inductor called an auto-transformer or several inductors. The main purpose of audio-frequency (AF) and radio-frequency (RF) transformers is to *isolate* the dc voltages of adjacent stages, while permitting the transfer of the alternating current (ac) signal. The *turns ratio* between the windings is important in the *impedance matching* and *voltage gain* of adjacent stages. Impedance is the total opposition offered to the flow of alternating current; adjacent stages should have similar impedances for the best power transfer.

The output voltage of an ideal transformer is directly proportional to the ratio of the number of turns in the output winding to those in the input winding. The main purpose of *power transformers* is to step-up or step-down ac voltages. The turns ratio between the windings is directly proportional to the ac voltages across the windings. These voltage ratios and *power-handling capabilities* in watts must be tabulated in the parts list.

The inductance and *coupling* of the two windings of an RF or intermediate-frequency (IF) transformer (Fig. 4-1e) may be changed by adjusting the position of the *ferrite core*. Turning an adjusting screw moves the ferrite core in or out of a coil. For more details, see Section 5-2.

Relays also serve to isolate circuits. When enough energy is supplied to the winding, the *armature* is magnetically attracted to the core of the winding. Movement of the armature causes the electrical *contacts* to open or close, thereby affecting the operation of an external circuit connected to these contacts.

The winding or coil of a relay is usually symbolized by a rectangle or circle as in Figure 4-1g instead of a corkscrew line. The rectangle may contain a small symbol or a letter combination to specify the type of relay; refer to Appendix G for more details. The small circle that appears in the left-hand square sig-

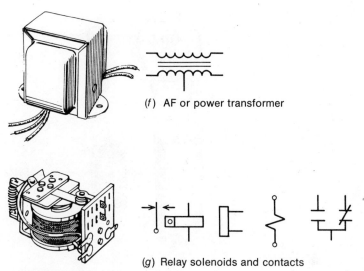

(f) AF or power transformer

(g) Relay solenoids and contacts

Figure 4-1 (*Continued*)

nifies that it is a fast-operating relay. A dot on one of the leads to a relay coil indicates the inner end of the winding.

The double set of short parallel lines in Figure 4-1*g* are known as transfer contacts. Contacts crossed with a diagonal line indicate *normally closed* contacts; when the relay is energized, normally closed contacts open and *normally open* contacts close. All contact symbols are drawn in their de-energized state.

4-3 Capacitors A *capacitor* consists of two conductors or metal plates separated by a *dielectric* (insulating) material. Its internal construction therefore relates directly to the symbols of Figure 4-2. It stores an electrical charge and also isolates dc voltages from affecting an adjacent stage while permitting the transfer of an ac signal.

There are two general types of capacitors, *polarized* and *nonpolarized*. Nonpolarized capacitors use *inert* insulating materials, such as air, paper, mica, and plastics, between the plates. The *positive polarity* of the dc circuit voltage may be connected to either terminal. Some cylindrically shaped nonpolarized capacitors (Fig. 4-2*a*) are marked with a black line near one end; the marked end terminal is usually connected to the chassis ground or point of lower potential.

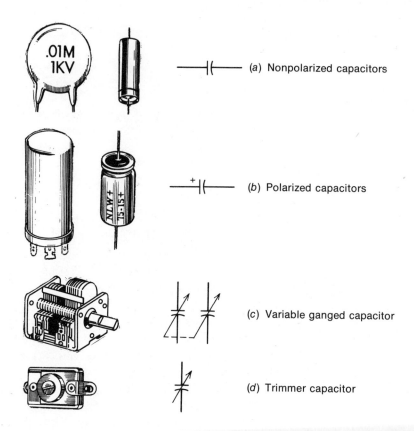

(a) Nonpolarized capacitors

(b) Polarized capacitors

(c) Variable ganged capacitor

(d) Trimmer capacitor

Figure 4-2 Capacitor symbols.

Polarized capacitors must be connected with the plus or red terminal connected to the more positive side of the circuit. The very thin insulating material of a polarized capacitor is formed through an *electrochemical process* and will be destroyed with *reversed polarity*.

Fixed value capacitors must be identified by their *capacitance* in *microfarads* (μF) or *picofarads* (pF) and the maximum or dc *working voltage* to which it is exposed.

The symbol for a *variable capacitor* includes a diagonal arrow, as in Figure 4-2c and d. The illustration in Figure 4-2c represents a *two-gang tuning* capacitor; its rated capacitance represents the capacitance when the moving plates completely engage the stationary plates. Both sets of moving plates in a two-gang capacitor are mounted on one shaft; the fact that they move together is symbolically indicated by the dashed line in Figure 4-2c. The bottom curved line of each symbol represents the moving plates and should always be connected to the *chassis ground*. Only the maximum capacitance in picofarads must be specified in the parts list.

Plates of the *trimmer* capacitor of Figure 4-2*d* are insulated from each other with thin sheets of mica. Capacitance of a trimmer is increased by turning the adjustment screw clockwise; this serves to bring the metal plates closer together. The trimmer capacitor is used in the final adjustment of a circuit and serves to compensate for slight differences in components.

4-4 A *resistor* is fabricated from a substance that is a poor con-
Resistors ductor of electricity, such as *carbon* or *nichrome wire,* and serves to oppose the flow of current in a circuit. The zigzag symbol for *resistance* in Figure 4-3*a* represents a difficult path for current flow.

Carbon resistors are presently manufactured in a cylindrical plastic shape enclosing the carbon material and are able to *dissipate* up to two *watts* of electrical power. Higher power resistors (Fig. 4-3*b*) must be fabricated from resistance wire encased in a *ceramic* material.

Carbon resistors are *color-coded* to specify their resistance in *ohms* (see Appendix C) and *tolerance*. The resistance value of wire-wound resistors is usually stamped on the body of the resistor.

Figure 4-3*c* illustrates an adjustable *tapped* resistor; the *clamp contact* may be moved during the final testing of a circuit to obtain optimum characteristics. When the center terminal is permanent, the symbol connection should be indicated with a dot; otherwise, an arrow is shown. Resistance of *trimmer resistors* (Fig. 4-3*d* and *e*) is changed with a screwdriver.

The resistance ratio of a *potentiometer* for panel mounting shown in Figure 4-3*f* is simply changed by rotation of the shaft. Figure 4-3*g* illustrates a *multiturn* potentiometer which is used in circuits where adjustments of greater precision must be made.

When only the center terminal and one end terminal is connected in the circuit, it is known as a *rheostat*. When used as a rheostat, it is good practice to connect the center terminal also to the remaining end terminal.

Parts list specifications of all resistance devices must include

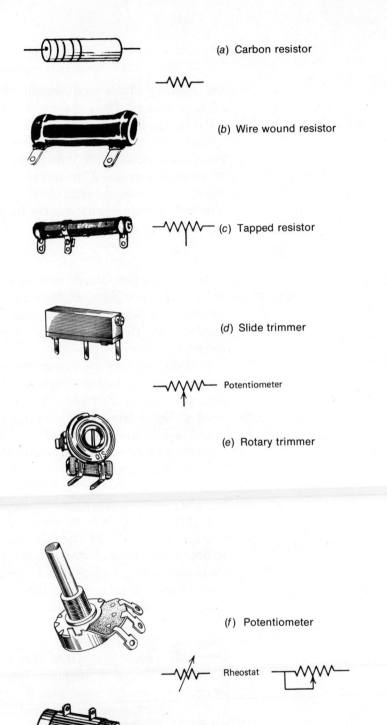

(a) Carbon resistor

(b) Wire wound resistor

(c) Tapped resistor

(d) Slide trimmer

Potentiometer

(e) Rotary trimmer

(f) Potentiometer

Rheostat

(g) Multiturn potentiometer

Figure 4-3 Resistance symbols.

the total resistance in ohms and their *power-dissipation capability* in watts.

4-5 All electronic circuits must contain a source of potential or a
Batteries and *power supply* to function. Portable electronic equipment
Cells usually is equipped with flashlight or *transistor batteries* as shown in Figure 4-4.

A *cell* contains two dissimilar plates or *electrodes* separated by a moist chemical mixture known as the *electrolyte; chemical energy* is converted to electrical energy when the cell is connected into a circuit. The symbol for a single cell (Fig. 4-4*a*) indicates the presence of the two dissimilar plates with a long thin line for the *positive terminal* and a short, thick line for the *negative terminal*. A *battery* consists of two or more cells connected in series, parallel or a series-parallel combination.

Size AA, C, and D cells all have a nominal potential of 1.5 *volts;* however, the D cell has the ability to furnish more power than the C or the AA penlight cell. When more potential is required, single cells are connected positive to negative terminal in a string or *in series* to form a higher voltage battery. Figure 4-4*b* illustrates three cells connected in series to obtain a total potential of 4.5 volts. It is not necessary to draw a symbol for each cell when many cells are connected in series; a dashed line may be used as illustrated in Figure 4-4*c*. The 9 volt transistor battery contains six small cells connected in series.

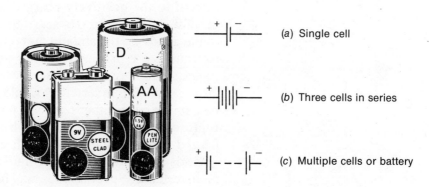

(a) Single cell

(b) Three cells in series

(c) Multiple cells or battery

Figure 4-4 Cell and battery symbols.

4-6
Switches

A *switch* is an *electromechanical* device used to open or complete an electrical circuit. The *toggle-type switch* of Figure 4-5a permits a faster separation of the contacts than the *slide-type switch* of Figure 4-5b. *Single-pole* (SP) switches are used to connect or disconnect a single-wire circuit while *double-pole* (DP) switches are used in two-wire circuits. When a *single-throw* (ST) switch is opened, a single circuit is disconnected; however, most *double-throw* (DT) switches can operate either of two circuits.

Push button or *momentary contact switches,* illustrated in Figure 4-5c, permit momentary opening or closing of a circuit. They are available with *multiple contacts* and *latch-down mechanisms.*

The symbol for a *rotary switch* (Fig. 4-5d) indicates the view from the back side of the panel, and the *operational sequence* is drawn in the clockwise direction. When a complicated *multiple deck* or *wafer* rotary switch is drawn, a *function table* is included, indicating which numbered terminals are connected for each switch position. Rotary switches are also specified as *shorting* or *nonshorting.* In the nonshorting type, no two adjacent contacts touch the *wiper blade* at the same instant.

Parts list specifications of switches should include the maximum permissible voltage and current that may be applied to the device.

4-7
Electron
Tubes

The *electron tube* consists of an *evacuated* glass or metal *envelope,* indicated symbolically with a circle, containing *cathode, anode,* and *grid* electrodes. The cathode emits *electrons* when heated by the *filament* or *heater.* The emitted electrons are attracted to the positively charged anode or *plate.* One or more meshlike grid electrodes serve to control the flow of electrons to the anode by the amount and polarity of their potential. Electron tubes are primarily voltage amplifying devices.

The *heater* symbol is the inverted V at the bottom of the circle; a double inverted V would indicate a heater that could be connected to either 6.3 or 12.6 volts. The inverted L symbol represents the cathode. Some tubes do not contain an *indirectly heated* cathode; the heater itself emits the electrons.

Grids are shown as dashed lines, while the anode is drawn as an inverted T. Several tubes may be contained in one envelope,

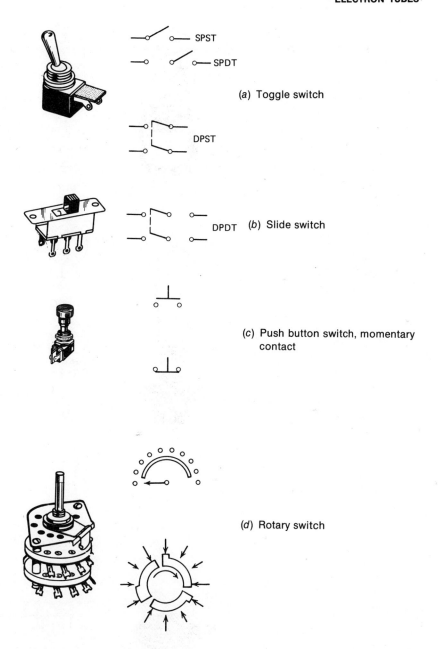

SPST

SPDT

(a) Toggle switch

DPST

DPDT   (b) Slide switch

(c) Push button switch, momentary
contact

(d) Rotary switch

Figure 4-5 Switch symbols.

as in Figure 4-6g, or the electrodes may be contained in incomplete circles.

Electron tubes are generally categorized by the number of

ELECTRONIC COMPONENT SYMBOLS

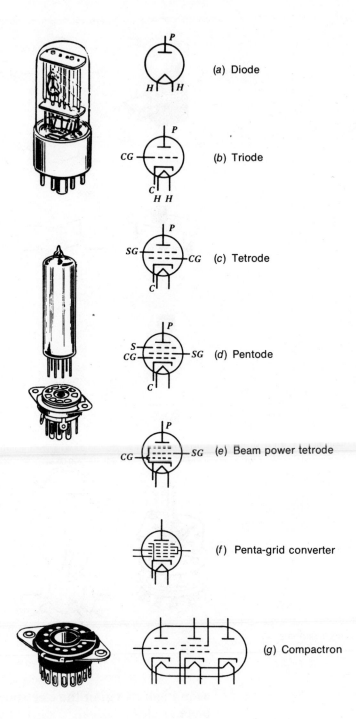

(a) Diode

(b) Triode

(c) Tetrode

(d) Pentode

(e) Beam power tetrode

(f) Penta-grid converter

(g) Compactron

Figure 4-6 Electron tube symbols.

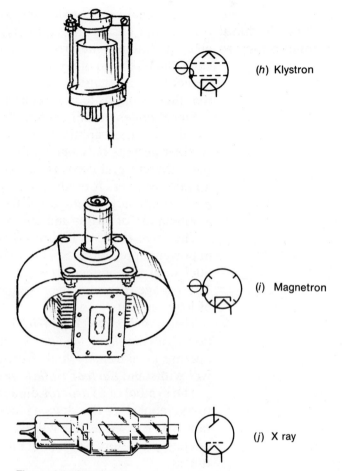

(h) Klystron

(i) Magnetron

(j) X ray

Figure 4-6 (*Continued*)

electrodes excluding the heater electrode. For instance, a *tetrode* (Fig. 4-6*c*) contains a cathode, two grids, and a plate.

It is usually desirable to add the pin numbers to the tube symbol. Pin designations are numbered clockwise from the *key* or *vacant pin* position, assuming that the viewpoint is from beneath the chassis. Pin numbers, specifications, and operating characteristics are obtainable from any tube manufacturer's data book. All manufacturers use the same system and tube numbers for tubes of identical characteristics. The left-hand numerals of a tube number indicate the approximate heater voltage; for example, a 6BA5 tube requires a 6.3 heater voltage; a 12AV6 requires a 12.6 v heater voltage and so on.

4-8
Two-Terminal
Semiconductors

Two-terminal *semiconductors* are generally known as *diodes;* they permit current flow in only one direction and under specific conditions. The arrow of the symbol in Figure 4-7a indicates the direction of *conventional current* flow from a positive potential to a negative or less positive potential. *Electron current* flow is opposite in direction to conventional current flow.

Since diodes permit current flow in only one direction, they usually find their application as *rectifiers* and *demodulators.* A rectifier permits only one half-*cycle* of an alternating current to pass; therefore, it converts ac to dc and finds its application in a radio receiver. A diode and a capacitor serve to separate the audio-frequency intelligence from the *modulated carrier* or combination of radio- and audio-frequency waves.

The *zener diode* in Figure 4-7b is used as a *voltage regulator* in power supply circuits. It is a special diode that does not conduct until the applied voltage reaches a *breakdown* voltage. The voltage then remains essentially constant and independent of load current.

The *tunnel diode* (Fig. 4-7c) is a special diode that can function as an *amplifier, oscillator,* or fast *switching device.* It can operate at extremely high frequencies and temperatures and will withstand *nuclear radiation* effects.

The symbol of a *varactor* diode (Fig. 4-7d) includes the capacitor symbol since this device includes a *voltage-sensitive* capacitance. It is used as the *voltage-tuned* capacitance in high-frequency oscillators, such as in the new *click-dial* UHF television tuners.

The *photodiode* or *solar cell* (Fig. 4-7g) is a *photovoltaic* device since it will convert light energy to electrical energy. Note that the arrows drawn adjacent to the diode symbol indicate the presence of light energy. Another type of photodiode is *photoconductive* because the resistance of the diode is reduced upon application of light energy. Photodiodes are used in *counting circuits, color-matching* devices, photographic exposure meters, door openers, and *optoelectronic* relays.

The *diac* is a two-terminal ac switching semiconductor. As noted from the arrows in the symbol in Figure 4-7e, this device can be made to conduct in either direction. Diacs are used in systems for motor speed control, heating controls, and light-dimming applications.

The *light-emitting diode* of Figure 4-7h is the newest develop-

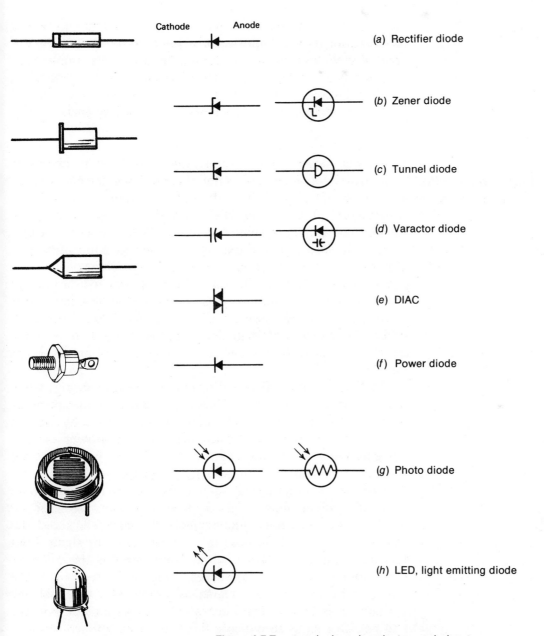

Cathode   Anode

(a) Rectifier diode

(b) Zener diode

(c) Tunnel diode

(d) Varactor diode

(e) DIAC

(f) Power diode

(g) Photo diode

(h) LED, light emitting diode

Figure 4-7 Two-terminal semiconductor symbols.

ment in two-terminal semiconductors. The symbol shows that its operation is the opposite of a photodiode in that light is emitted when current flows through the diode. Its simplest application is as a *pilot lamp* since it consumes less energy than an incandescent lamp. In multiple form, it serves as the *digital readout* in calculators and electrical measuring instruments.

4-9 Multiple-Terminal Semiconductors

In general, three-terminal semiconductors differ in operation from two-terminal devices in that the additional terminal serves to control the current flow through the semiconductor. The *silicon-controlled rectifier, triac,* and *diac* are all members of the *thyrister* family of semiconductors. They are mostly heavy current devices that are extensively used in applications involving power rectification, control, and switching.

The silicon-controlled rectifier (SCR) is similar to a relay in that a small current permits a heavy current to flow in another circuit; however the two circuits are not isolated from each other as in a relay. A small signal applied to the *gate* terminal causes current to flow between the cathode and anode terminals.

The SCR is a *unidirectional reverse-blocking thyrister,* while the *triac* of Fig. 4-8*b* is a *bidirectional* device that can block voltages of either polarity. This ability is suggested by the two arrows in the symbol. Triacs are used in power control systems, such as motor control, heating, arc welding, light dimmers, and other power-switching applications.

A *transistor* is an *active* semiconductor with at least three terminals, which may be used as an amplifier, detector, or switch. With reference to the symbols of Figure 4-8*c* and *d*, the vertical line is called the *base* (*B*) and is usually the signal input terminal. The diagonal line within the symbol is the *collector* (C) terminal and most often serves as the output terminal of the device. The *emitter* (E) terminal is shown as a diagonal line containing an arrow. The illustrated symbols include positive and negative signs to indicate the polarity of applied voltage.

There are two basic types of *bipolar* transistors (BJT), the *npn* and the *pnp,* which are determined by the arrangement and *doping* of their semiconductor materials. In a *pnp* transistor symbol, the emitter arrow points *toward* the base; in an *npn* transistor, the arrow points *away from* the base. The arrow

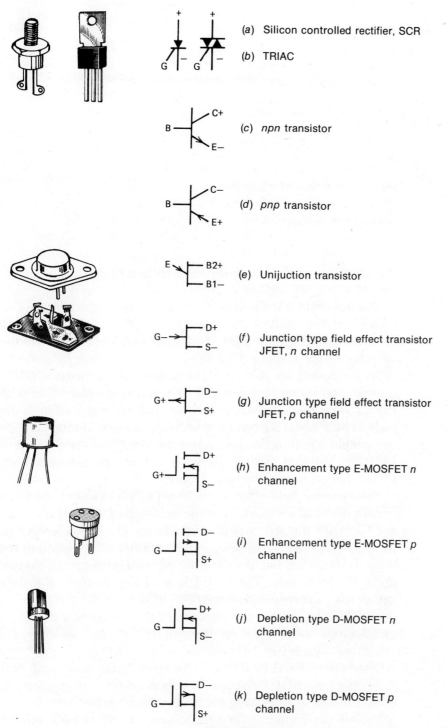

(a) Silicon controlled rectifier, SCR

(b) TRIAC

(c) *npn* transistor

(d) *pnp* transistor

(e) Unijuction transistor

(f) Junction type field effect transistor JFET, *n* channel

(g) Junction type field effect transistor JFET, *p* channel

(h) Enhancement type E-MOSFET *n* channel

(i) Enhancement type E-MOSFET *p* channel

(j) Depletion type D-MOSFET *n* channel

(k) Depletion type D-MOSFET *p* channel

Figure 4-8 Multiple-terminal semiconductor symbols.

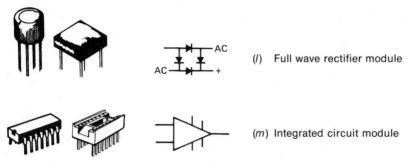

(*l*) Full wave rectifier module

(*m*) Integrated circuit module

Figure 4-8 (*Continued*)

direction shows the *conventional* direction of current in the active or saturated mode.

The common transistor contains a *junction* between the collector and base and one between the emitter and base. Therefore, the two diagonal lines should not touch each other in the drawing of the symbol.

As suggested by its name, the *unijunction* transistor (UJT) contains only one junction (Fig. 4-8*e*). The diagonal line with the arrow is again called the emitter, and the vertical line is the base. The *B*1 and *B*2 base connections are not junctions; they are simple *ohmic* or welded contacts. Applications of the unijunction transistor include *oscillator* circuits, *pulse generators*, *phase detectors*, and *trigger* circuits.

The *junction field-effect transistor* (J FET) symbol shown in Figure 4-8*f* and *g* differs from the unijunction transistor in that the line with the arrow is drawn horizontally and is known as the *gate* (G). The signal to be amplified is usually applied to the gate. The vertical line is called the *channel* with *source* (S) and *drain* (D) terminals. The J FET is a *voltage-controlled* device rather than a *current-controlled* device like a bipolar transistor; it therefore has high *input* impedance characteristics similar to an electron tube. J FETs are commonly used in *chopper*, *switching*, and *mixer* circuits.

The symbol for the MOS FET or *metal oxide semiconductor field-effect transistor* contains a capacitor-type input gate terminal. A thin layer of an insulating material exists between the gate and the channel; for this reason the MOS FET has an extremely high input impedance. MOS FETs are one of the least *temperature-dependent* semiconductors available today. Their most important application is in *digital computer switching* circuits.

4-10
Accessory
Devices
Electronic circuits can be damaged from the effects of *over-loading* or by components that have become defective. Protection of circuits may be accomplished through the inclusion of *fuses* or *circuit breakers,* as pictorially and symbolically illustrated in Figure 4-9m and *p.* A fuse contains a section of conductor that melts when the current through it exceeds a rated value for a definite period of time. A circuit breaker is an electromagnetic device that automatically opens a circuit when the current exceeds the rating of the circuit breaker. The parts list must include the voltage and current rating of these devices.

The *dot system* to indicate connections in electronic schematic diagrams is generally accepted. *Crossovers* of conductors in a schematic diagram are usually made at 90° and do not contain the dot. If one or more conductors are combined in a *shielded* cable, then a circle or ellipse should be drawn to include the conductors (Fig. 4-9*b*). The shielding braid of a *coaxial* cable is also used as a conductor. Figure 4-9*a* illustrates how a multiple conductor cable without shielding is drawn.

An almost endless variety of connecting devices exist for easy coupling of wires and cables to electronic equipment. A removable connector usually consists of one or more *sockets* (female part) and an equal number of pins on the *plug* (male part), which mate as in Figure 4-9*c*. Most multiple connectors are *polarized* by arranging the pins in a nonsymmetrical pattern or by using pins with different diameters.

Most electronic equipment is connected to some form of *transducer* which converts an electrical or mechanical input signal into an output signal of a form compatible with the electronic circuit. Pictorial and symbolic illustrations of many such input devices are shown in Figure 4-10.

Similarly, all electronic equipment must be terminated with a transducer to convert the electrical output to sound, light, or mechanical energy as illustrated in Figure 4-11. *Acoustic devices,* such as *headphones* and *speakers,* provide an audible response to an electric current that is circulated through their windings. Television *picture tubes, cathode-ray tubes, incandescent lamps, gas tubes,* and *digital-readout tubes* are examples of output devices that convert electrical energy to light energy.

The symbol for an electrical meter is always a circle that contains the acronym indicating the type of measuring instrument. Often the *full-scale* value that the meter is capable of measuring is added below the circle.

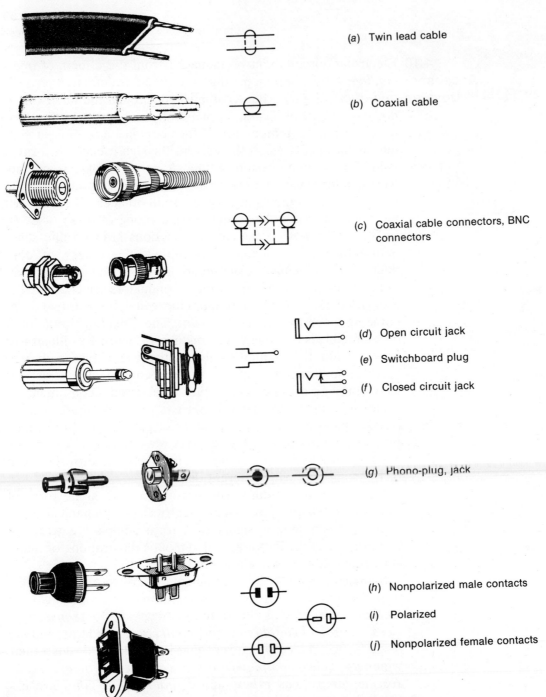

(a) Twin lead cable

(b) Coaxial cable

(c) Coaxial cable connectors, BNC connectors

(d) Open circuit jack

(e) Switchboard plug

(f) Closed circuit jack

(g) Phono-plug, jack

(h) Nonpolarized male contacts

(i) Polarized

(j) Nonpolarized female contacts

Figure 4-9 Cable and connector symbols.

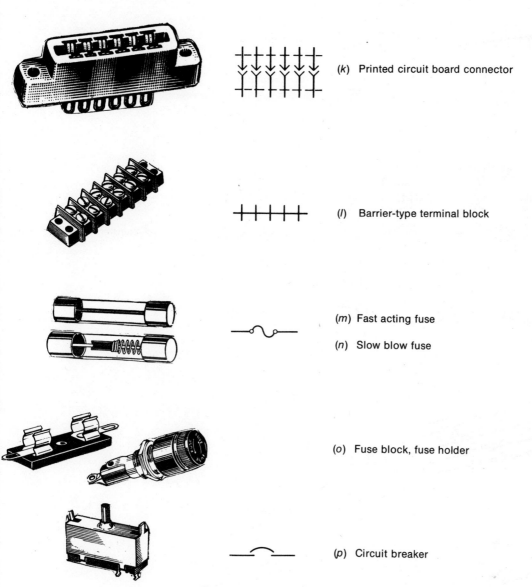

(k) Printed circuit board connector

(l) Barrier-type terminal block

(m) Fast acting fuse

(n) Slow blow fuse

(o) Fuse block, fuse holder

(p) Circuit breaker

Figure 4-9 (*Continued*)

4-11 Size of Symbols The size of an electronic symbol is not related to the size of the actual component. Within an electronic diagram, each similar symbol must be identical with regard to size and proportion. For example, all transistor symbols should be contained in the same diameter circle envelope. Templates with a variety of cut-outs or appliqués can be obtained to suit most of the requirements of electronic diagrams; their use will obviously produce symbols that are reasonably proportionate to each other in

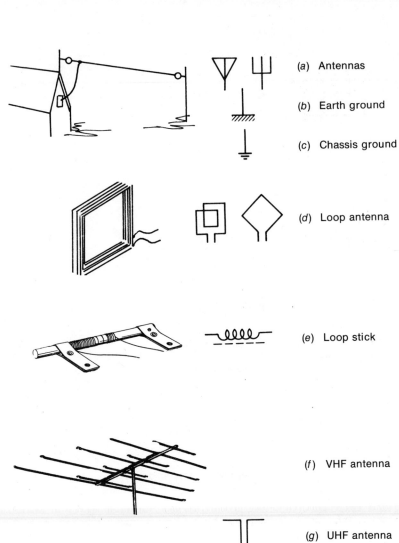

(a) Antennas

(b) Earth ground

(c) Chassis ground

(d) Loop antenna

(e) Loop stick

(f) VHF antenna

(g) UHF antenna

(h) Crystal microphone

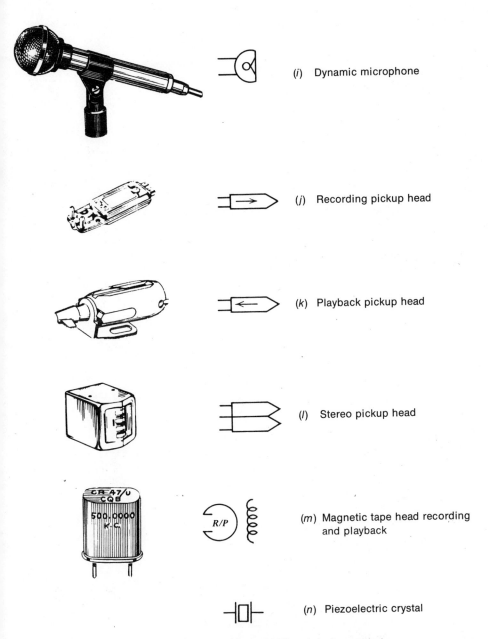

(i) Dynamic microphone

(j) Recording pickup head

(k) Playback pickup head

(l) Stereo pickup head

(m) Magnetic tape head recording and playback

(n) Piezoelectric crystal

Figure 4-10 Input device symbols.

ELECTRONIC COMPONENT SYMBOLS

(*a*) Receiver earphone

(*b*) Stereo headset

(*c*) Loudspeaker

(*d*) Milliameter

(*e*) Voltmeter

Figure 4-11 Output device symbols.

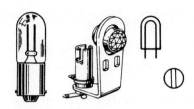

(*f*) Indicating lamp

(*g*) Jeweled signal light

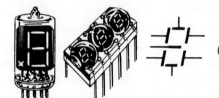

(*h*) Digital readouts

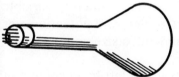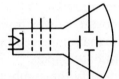

(*i*) Cathode-ray tube

Figure 4-11 (*Continued*)

size. It is suggested that the size of symbols be drawn within the following limits:

1. Resistor height, between 3 mm ($^1/_8$ in.) and 6 mm ($^1/_4$ in.).
2. Resistor length, at least three points on each side.
3. Inductor loop diameter, between 2 mm ($^3/_{32}$ in.) and 6 mm ($^1/_4$ in.).
4. Capacitor plate separation, 1.0 mm ($^1/_{32}$ in.) and 2.5 mm ($^3/_{32}$ in.).
5. Capacitor plate lengths, between 4 mm ($^5/_{32}$ in.) and 9 mm ($^{11}/_{32}$ in.).
6. Transistor or tube envelope, between 10 mm ($^{13}/_{32}$ in.) and 20 mm ($^{25}/_{32}$ in.)

Summary 1. Electronic symbols are representations used to typify electronic components.
2. Electronic symbols are a technical shorthand.
3. Electronic symbols have been standardized and accepted by

industry and the military so that universal communication is possible.

4. Symbols of electronic devices containing comparable construction details usually have some similarities in their symbol design.

5. Since the shape of an electronic symbol is related to the function and construction of the device, knowledge of component function and construction is also important as an aid in remembering symbol shapes.

6. When an electronic component is used more than once in a circuit, the symbol must be consistent in shape and size.

7. Use of electronic symbol templates or appliqués will produce consistent symbol shapes and sizes in addition to reducing the time needed to draw the symbols.

Problems 4-1 Make freehand drawings of a resistor, inductor, variable capacitor, rheostat, power transformer, and potentiometer symbols to the maximum recommended size on metric rectangular coordinate paper.

4-2 Completely identify the symbols in the left-hand column of Figure 4-12.

4-3 Draw the proper symbols corresponding to the electronic devices in the right-hand column of Figure 4-12.

4-4 Study the symbols of Figure 4-13; what device is each symbol intended to represent? Redraw each symbol in the correct form and proportion.

4-5 Draw the symbol of a compactron containing a triode, tetrode, and a beam power tube.

4-6 Identify all the electronic components in the wiring diagram in Figure 4-14. Prepare a parts list including typical component specifications.

4-7 Determine the complete specifications of the following color-coded resistors.
   a. Brown, red, brown, and gold.
   b. Violet, green, yellow, and brown.
   c. Orange, blue, green, and silver.
   d. Gray, green, red, and none.
   e. Red, brown, blue, and silver.

4-8 Determine the possible range of resistance for each of the resistors in Problem 4-7.

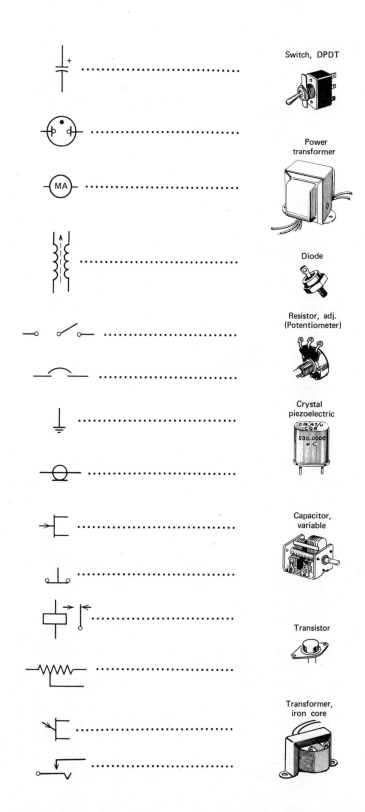

Figure 4-12 Symbol problem.

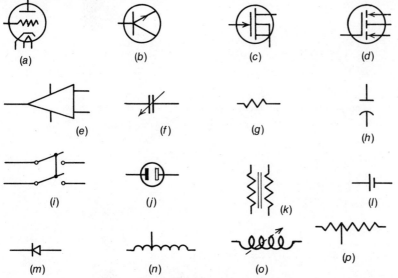

Figure **4-13** Improper symbols for use in Problem 4-3.

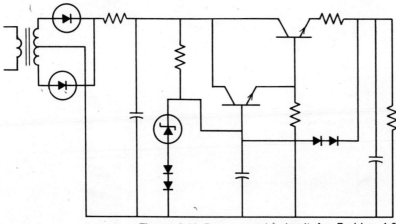

Figure **4-14** Power supply circuit for Problem 4-6.

4-9 Find the color codes of the following resistors.
  a. 470 ± 5% ohms
  b. 2.5 ± 20% megohms
  c. 0.39 ± 10% ohms
  d. 6.8 ± 10% kilohms
  e. 0.15 ± 5% megohms

Now return to the objectives and self-evaluation questions at the beginning of this chapter and see how well you can answer them. If you cannot answer certain questions, place a check next to these and review appropriate parts of the text to find the answer.

References **Charles J. Baer,** *Electrical and Electronics Drawing,* Third Edition, McGraw-Hill, New York, 1973, pp. 40–61, 365–367.

Nicholas M. Raskhodoff, *Electronic Drafting and Design,* Second Edition, Prentice-Hall, Englewood Cliffs, N.J., 1966, pp. 78–139.

George Shiers, *Electronic Drafting,* Prentice-Hall, Englewood Cliffs, N.J., 1962, pp. 203–237.

Chapter 5 Basic Circuits

Instructional Objectives

1. To become more familiar with electronic symbols through circuit application.
2. To understand the purpose of schematic diagrams.
3. To develop a basic electronic vocabulary.
4. To learn the basic elements needed in all electronic circuits.
5. To become aware of the basic circuits which make up a complete electronic device.
6. To make it possible for the student to trace a circuit using a schematic diagram.

Self-Evaluation Questions Test your prior knowledge of the information in this chapter by answering the following questions. Watch for the answers as you read the chapter. Your final evaluation of whether you understand the material is measured by your ability to answer these questions. When you have completed the chapter, return to this section and answer the questions again.

1. What is the purpose of a schematic diagram?
2. Define the word *stage*.
3. In what location should the input to a circuit be drawn?
4. What determines the position of a stage within a schematic diagram?
5. How are stages of a multistage circuit connected to each other?
6. Where should the loudspeaker symbol in a radio receiver circuit be drawn?
7. How is a polarized capacitor drawn in a circuit path?
8. What basic devices are needed in a power supply circuit?
9. What components are different in RF and AF amplifiers?
10. Where are auxiliary circuits normally placed in a schematic diagram?

**5-1
Power
Supply
Circuits**
Voltage of 115 volts ac, available at the *duplex wall receptacle* in every home, varies continuously in value and periodically reverses in polarity. Most electronic circuits require a dc supply voltage that is constant in value and polarity. An electron tube or semiconductor diode allows current to flow in only one direction; therefore, it permits only voltages of the same polarity to flow, as shown in the *half-wave* shape of Figure 5-1a. The dashed portion of the wave shape will not be passed to the load.

Full-wave rectification is obtained if a circuit is added to rectify the alternate half-cycles. Instead of using another transformer, a center-tapped transformer secondary and two diodes are connected as in Figure 5-1b. Note that the output waveform is smoother; full-wave rectification is more efficient and requires less *filtering*. As discussed in Section 4-5, the zener diode, D_3, maintains the output voltage essentially constant. Additional voltage produced will be absorbed by the R_1 *bleeder* resistor.

The *voltage-doubler* circuit of Figure 5-1c is capable of delivering approximately twice the dc voltage supplied by the full-wave circuit. In low-power circuits, it has the advantage of not requiring a transformer. As in the operation of the full-wave rectifier, each diode conducts only when its anode is on the positive cycle and then charges its series capacitor. Since the two capacitors are connected in series across the output, their charges add to double the voltage. Figure 5-1c also includes a *filter choke* and capacitor to reduce the *ripple* or pulsations in the dc output.

**5-2
Amplifier
Circuits**
Many electronic circuits are basically *amplifier* circuits. In the most popular (common emitter) transistor-amplifier circuit, the *input* signal is applied to the *base* terminal and the amplified *output* is obtained at the *collector* terminal (Fig. 5-2a). Since the *emitter* terminal is common to both the input and output of the circuit, it is generally referred to as the *common-emitter* (CE) amplifier circuit. In comparison with an electron tube circuit, changes in the voltage applied to the *grid* cause an amplified change in the *plate* circuit; the *cathode* is common to both the input and output circuits. (Note that electron tubes are voltage-

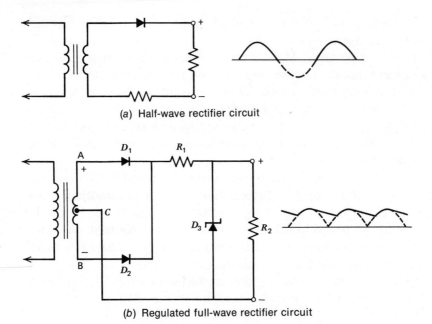

(a) Half-wave rectifier circuit

(b) Regulated full-wave rectifier circuit

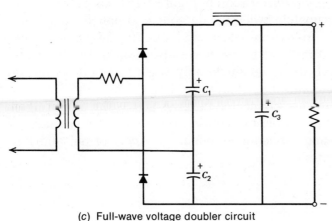

(c) Full-wave voltage doubler circuit

Figure 5-1 Power supply circuits.

amplifying devices instead of current amplifiers as are transistors.)

In another configuration, the output is obtained from the emitter terminal, called the *common-collector* circuit (CC), shown in Figure 5-2b. The circuit *voltage gain* is now reduced to less than one, although the *current gain* remains high. The common-collector amplifiers' *input impedance* is considerably higher than that of the CE amplifier; this characteristic makes

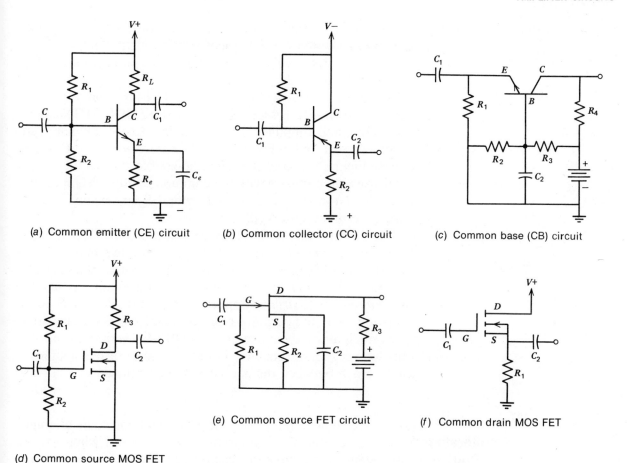

(a) Common emitter (CE) circuit

(b) Common collector (CC) circuit

(c) Common base (CB) circuit

(e) Common source FET circuit

(f) Common drain MOS FET

(d) Common source MOS FET

Figure 5-2 Single-stage amplifier circuits.

the CC circuit useful in *coupling* high-impedance input devices such as the *crystal microphone*. The output of this amplifier almost matches the low impedance of *coaxial cables*.

As indicated in Figure 5-2c, the emitter terminal is the input terminal of the *common-base* amplifier. Electromagnetic input devices are easily coupled to this amplifier because it has the lowest input impedance of the three amplifier circuits. It is the least popular amplifier circuit because of its lack of current gain and its very high output impedance.

The CE amplifier (Fig. 5-2a) is the only one of the three circuits having *both* voltage gain and current gain, resulting in the highest power gain. All three amplifier circuits are the same for *npn* or *pnp* transistors; however, the battery polarities are reversed. Note that the collector terminals of the *npn* transistors in Figure 5-2*a* and *c* are connected to a positive potential. On the other hand, the collector is at a negative potential in the *pnp* circuit of Figure 5-2*b*.

The *junction field-effect transistor* (J FET) and the *metal oxide semiconductor field-effect transistor* (MOS FET) may be connected into amplifier circuits similar to the bipolar transistor as illustrated in Figure 5-2. The *common-source* FET circuit of Figure 5-2*e* and the common-source MOS FET circuit of Figure 5-2*d* compare with the common-emitter circuit of Figure 5-2*a*. Note that the *source terminal* S corresponds to the emitter terminal of a bipolar transistor.

The J FET and MOS FET are *voltage-controlled* devices rather than *current-controlled* devices like the bipolar transistors. For this reason they have high input impedance like electron tubes and have become very important because of their *integrated circuit* applications. They are also more radiation- and temperature-resistant than bipolar transistors. The principal disadvantage of the MOS FET is that it is unsuitable for very high-frequency circuits because of the significant capacitance between the gate and drain terminals.

5-3 Interstage Coupling Most electronic applications use more than one stage of amplification. Several precautions must be observed when two or more stages are *coupled* to each other. The output impedance of the first amplifier stage should match the input impedance of the second stage for maximum power transfer. It is also important that the dc output potential is isolated from the input of the following stage. There are three basic methods used to couple transistor stages:

1. Transformer coupling.
2. Resistance-capacitance coupling (RC).
3. Direct coupling (DC).

When the signal to be amplified is at a high frequency, such as at radio frequency (RF) or intermediate frequencies (IF), the transformer should have an air or ferrite core to reduce losses. Since the windings of a transformer are only *magnetically coupled,* the dc potentials of adjacent stages will be isolated. The primary (input) and secondary (output) windings are designed to match the impedances of the coupled stages.

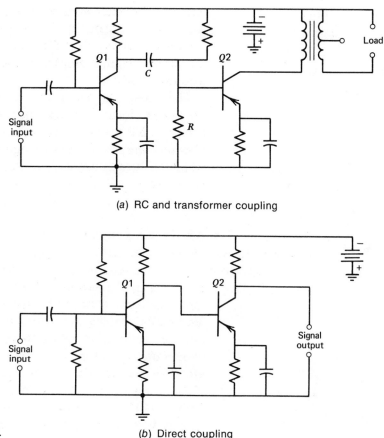

(a) RC and transformer coupling

(b) Direct coupling

Figure 5-3 Multistage amplifier circuits.

The *output transformer* shown in the circuit in Figure 5-3*a* is wound on a more efficient laminated iron core; this core is permissible on low or *audio frequencies*. The primary winding has an impedance to match the output impedance of the second common-emitter amplifier stage, while the secondary impedance is very low to match the speakers' impedance.

Interstage transformer-coupled amplifiers do not ordinarily possess as flat a *frequency response* as is obtainable through the use of *resistance-capacitance* (RC) *coupling*. RC coupling is also less expensive than transformer coupling; however, it has more power loss.

Figure 5-3*b* shows *direct-coupling* (DC) of amplifier stages in that the Q_1 collector is directly connected to the base of the Q_2 transistor. This type of coupling is used in many IC amplifiers because the other methods require bulky and expensive capacitors and transformers. Transformer and RC-coupled amplifiers are only effective in the amplification of an ac signal, whereas the direct coupled amplifier response can be extended to zero frequency or dc.

The complete AM radio receiver in Figure 5-5 should be checked to locate the amplifier stages. Note that the CE amplifier Q_2 is RF transformer-coupled at both its input and output. Furthermore, the *detector* stage is RC-coupled to transistor Q_3, which in turn is transformer-coupled to the Q_4–Q_5 *push-pull* amplifier.

5-4
Oscillator
Circuits

An *oscillator* is a circuit that converts dc into ac energy of some *periodic waveform* at a specific frequency. The primary difference between an amplifier circuit and an oscillator circuit is that part of the output current is returned *in phase* with the input current in order to sustain oscillations. This in-phase feedback is known as *positive* or *regenerative feedback*. When one ac wave is in phase with another, their *amplitudes* or instantaneous values are of the same polarity and may be combined by simple addition to a greater instantaneous value.

The simplest oscillator circuit is shown in Figure 5-4a; it is known as the *tickler coil* or *regenerative* oscillator. The resonant inductive-capacitive (LC) circuit induces a periodic alternating voltage in the tickler coil; this induced ac is fed back in phase via C_1 to the base input of the transistor. The values of inductance L and capacitance C, connected in parallel, determine its *resonant frequency*. At the resonant frequency, the *reactance,* or opposition to ac flow, of the inductor and capacitor are equal and opposite in direction and cancel each other. This condition produces maximum impedance and therefore maximum voltage across inductor L; a maximum voltage is then induced in the tickler coil L_1.

Another oscillator used in radio broadcast receivers is the *Hartley* circuit (Fig. 5-4b). Part of the voltage developed in the L-C-L_1 resonant circuit is fed back to the transistors' base input via the coupling capacitor C_2.

Regenerative feedback from the collector to the emitter is achieved by the capacitors in the *Colpitts* oscillator (Fig. 5-4c). Observe that the capacitor string is effectively tapped instead of the inductor as in the Hartley circuit.

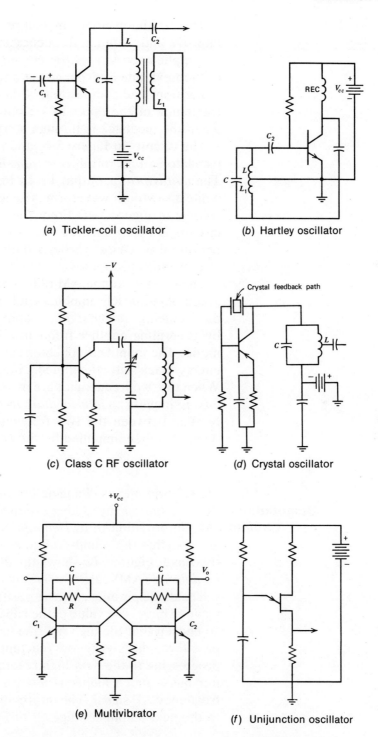

(a) Tickler-coil oscillator

(b) Hartley oscillator

(c) Class C RF oscillator

(d) Crystal oscillator

(e) Multivibrator

(f) Unijunction oscillator

Figure 5-4 Oscillator circuits.

The oscillator circuit in Figure 5-4d is *crystal-controlled* to improve and maintain frequency stability. An alternating voltage applied across a *quartz crystal* causes it to vibrate; this vibration exhibits a resonance at a frequency determined by the dimensions and composition of the crystal. The broad resonant frequency of the *tank* (LC) circuit is considerably improved when the feedback is through a crystal.

The circuits in Figure 5-4c and *f* represent *relaxation* or RC oscillators; they produce a *nonsinusoidal* output waveform. The *multivibrator* output *Vo* in Figure 5-4e is a square wave, while a sawtooth waveform is generated by the *unijunction oscillator* in Figure 5-4f. Frequency of oscillations in relaxation circuits is determined by the rate at which the capacitors are permitted to charge through their series resistors and by the size of the capacitors.

The Q_1 stage of the AM radio receiver circuit in Figure 5-5 is a combination RF amplifier and oscillator. The frequency of the incoming *carrier signal* is converted to 455 kHz (kilohertz) by generating another frequency in the transistors' oscillator mode. The amplified collector current in the inductor L feeds energy back into the tuned circuit to maintain oscillations. When the two frequencies are mixed together, the results of the *mixing* include an *intermediate frequency* (IF) which is the difference between the two frequencies. This intermediate frequency is then amplified by the Q_2 common-emitter stage.

5-5 Demodulator Circuits *Modulation* means changing the radio-frequency (RF) wave at the *transmitter* by adding sound-producing audio-frequency (AF) information to it. *Demodulation* means changing it back again so that the sound-producing information is obtained from the wave. Figure 5-6a shows the demodulation of an *amplitude modulated* (AM) carrier wave by a *detector* circuit.

The diode permits only the positive portion of the modulated wave to pass. The value of the capacitor (following the diode) is so chosen that the high intermediate frequency (IF) prefers to pass through its very low reactance path to ground instead of progressing to the next high reactance stage. The same capacitor has a much higher reactance or opposition to the lower frequency AF signal. The information at AF therefore proceeds to the next amplifier stage Q_3 of Figure 5-5.

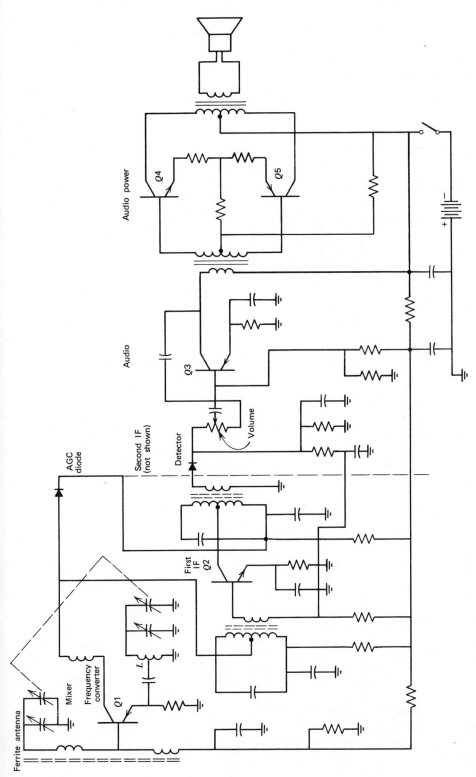

Figure 5-5 Complete AM radio receiver circuit.

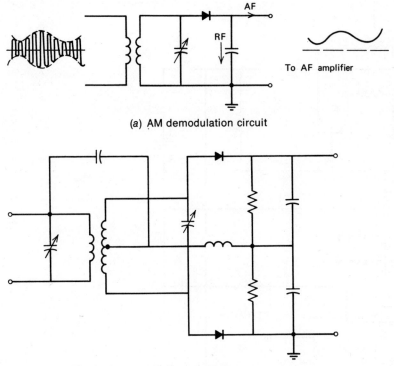

(a) AM demodulation circuit

(b) FM phase discriminator

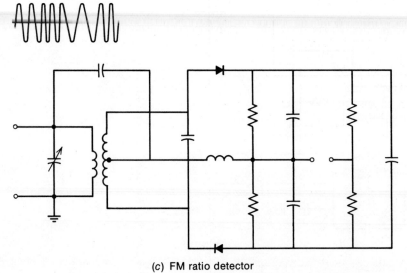

(c) FM ratio detector

Figure 5-6 Demodulator circuits.

In *frequency modulation* (FM), the *amplitude* or peak-to-peak height of the carrier wave remains steady, but the frequency changes in proportion to the frequency of the AF intelligence. Lightning and other sources of interfering *noise* are *limited* by the FM receiver. The FM demodulator circuit (Fig. 5-6*c*) includes a limiter circuit to remove spurious signals in addition to producing an output whose amplitude depends only on the *frequency deviation* of the input signal, which carries the intelligence.

The *phase discriminator* demodulator circuit (Fig. 5-6*b*) provides a greater output and is more *linear* than the *ratio detector* circuit of Figure 5-6*c* but must be preceded by a limiter stage. For circuit identification purposes, note that the FM demodulators contain two diodes instead of only one for AM demodulation. The ratio-detector circuit is further identified by its *reversed* diodes.

5-6 Filter Circuits Many electronic circuits simultaneously contain currents of different frequencies. This was discussed in connection with the demodulation circuits of Section 5-5. The small capacitor in the AM detector circuit of Figure 5-5 removes or *filters out* the intermediate frequency; it may therefore be called a *low-pass filter* (Fig. 5-7*a*).

High-pass and low-pass filters reject or pass certain frequencies as a result of the way the frequency affects their *reactance* or opposition to current flow. An inductor has more opposition to ac as the frequency increases. On the other hand, a capacitor opposes ac to a greater extent as the frequency decreases. In the low-pass filter circuits of Figure 5-7, the inductors are in *series* with the line and offer little opposition to low frequencies, whereas the *parallel-connected* capacitors act almost as a short circuit to high frequencies because of their low reactance at high frequencies.

Since the capacitors in a *high-pass* filter are in series with the line, they offer little opposition to high frequencies but considerable opposition to low frequencies. Similarly, the inductors act as short circuits to low frequencies (Fig. 5-7*d*, *e*, and *f*).

The *band-pass* (Fig. 5-7*g* and *h*) and *band-stop filters* (Fig. 5-7*i* and *j*) contain resonant circuits. As discussed in Section

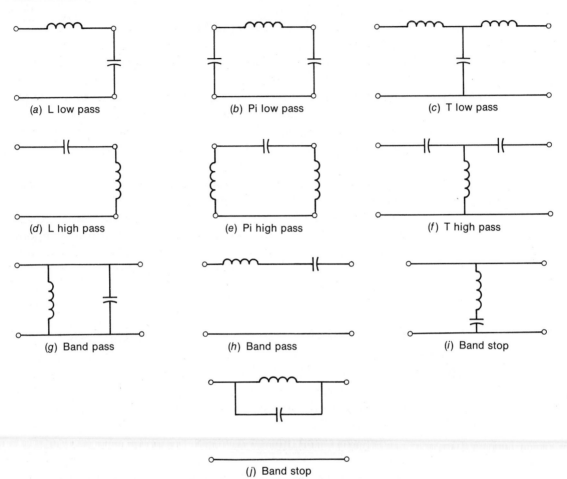

(a) L low pass (b) Pi low pass (c) T low pass

(d) L high pass (e) Pi high pass (f) T high pass

(g) Band pass (h) Band pass (i) Band stop

(j) Band stop

Figure 5-7 Filter circuits.

5-3, the condition of resonance exists at a definite frequency, depending on the values of inductance and capacitance.

The band-pass filter in Figure 5-7g contains a parallel resonant LC circuit across the line; since the *impedance* of a parallel LC circuit is *maximum* at resonance, the filter will not short currents at the resonant frequency to ground. When the resonant circuit has both components in series with the line (Fig. 5-7h), the circuit also passes only the series LC resonant frequency since at resonance a series LC circuit has minimum impedance.

When the resonant circuit connections are reversed as in Figures 5-7i and j, the filters automatically *reject* the resonant frequency band. For instance, at the resonant frequency of the series LC circuit of Figure 5-7*i*, the impedance at the resonant frequency is almost zero, shorting out all signals at the resonant frequency. Frequencies above and below resonance pass through the filter without much *attenuation* or loss.

5-7 Integrated Circuits An *integrated circuit* (IC) may be defined as an arrangement of active and passive semiconductor devices into a miniaturized circuit *module*. This development of solid-state technology has (1) reduced electronic circuit costs; (2) made the circuits more reliable; (3) in some instances, improved circuit performance and (4) reduced both size and operating power.

The *operational amplifier* (OP AMP) is one of many spectacular integrated circuit developments. The Fairchild 741 μA module contains 20 transistors and 11 resistors and is smaller in physical size than 3 conventional transistors. As shown in the circuit in Figure 5-8*a*, only 2 additional resistors and a power supply are needed for an amplifier circuit with a gain of at least 100.

The circuit in Figure 5-8*b* contains a revolutionary *voltage regulator* device in addition to the transformer, diode, and shunt capacitor of the standard half-wave rectifier circuit. This IC module maintains the output voltage constant and independent of load currents up to 1 ampere.

The 555 functional IC (Fig. 5-8*c*) is a highly stable device for generating accurate time delays and square or sawtooth oscillations. The module contains 27 transistors and 11 resistors within a volume of about 0.5 cm³. An external variable capacitor, *F*, controls the frequency of oscillation while an external rheostat, *W*, determines the *pulse width* of the output waveform.

The ICs shown in Figure 5-8*d* convert a time base from the 60 Hz power line to a *light-emitting diode* (LED) digit time display. The 60 pulses per second are divided by counter modules such as the 7490 to obtain one pulse per second. The 7447 *decoder-driver* IC then supplies energy to the seven segments of the LED numeric display in proper order. Every second a different combination of segments in lighted to display time in seconds from zero to nine numerically. Five similar units are needed to complete the hours, minutes, and seconds display,

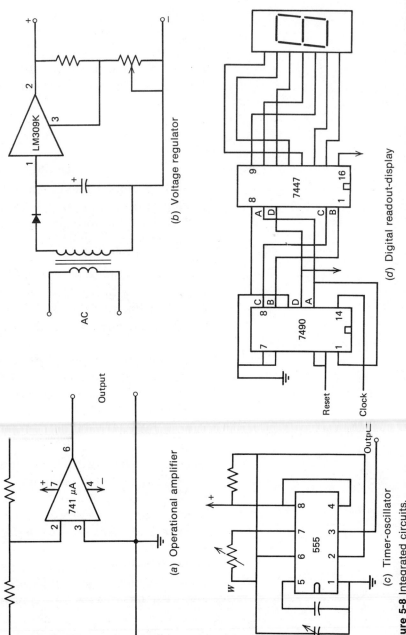

(a) Operational amplifier

(b) Voltage regulator

(c) Timer-oscillator

(d) Digital readout-display

Figure 5-8 Integrated circuits.

although later developments have combined sets of these small ICs into a single but larger IC.

Summary **1.** Solid-state components such as discrete active devices are combined with passive devices and a source of energy to form a functional circuit.

2. Most electronic devices contain stages with different operational characteristics such as amplifiers, filters, oscillators, and rectifiers.

3. A transistor may be connected variously in three ac amplifier circuits with different operating characteristics.

4. If a portion of an amplifier's output is fed back in phase with the input, the circuit functions as an oscillator.

5. All radio receiver circuits must contain a demodulator stage to separate the sound-producing intelligence from the transmitted radio signal.

6. Most electronic circuits must be provided with a dc source to function properly.

7. When only ac is available, dc is obtained through stages of rectification and filtering.

8. Requirements of the computer age have served to miniaturize and improve many electronic circuits through small-, medium-, and large-scale integration of components.

Problems **5-1** Identify each circuit in Figure 5-9; give your reasons for such identification.

5-2 Does the circuit in Figure 5-9a contain two CB, CC, or CE amplifiers; explain your answer.

5-3 What is the purpose of the piezoelectric crystal in Figure 5-9b?

5-4 If *npn* transistors having the same characteristics are substituted in Figure 5-9d, what other changes must be made?

5-5 What is the purpose of resistor *R* in the circuit in Figure 5-9e?

5-6 What is the function of (a) the choke and (b) the capacitors in the circuit in Figure 5-9f?

5-7 Identify each circuit in Figure 5-10 and give your reasons for each identification.

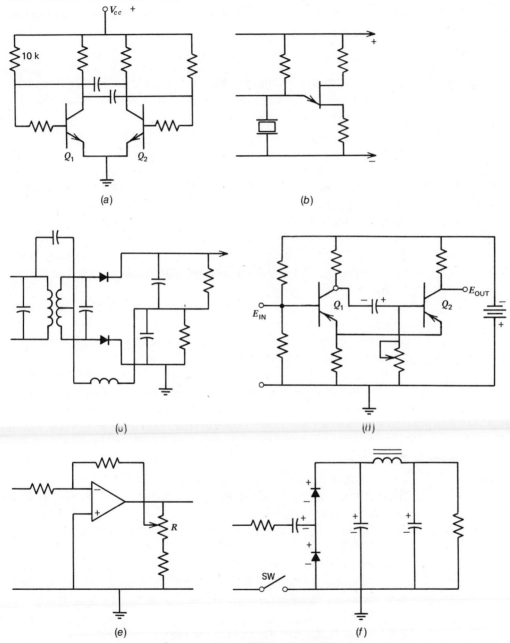

Figure 5-9 Circuit identification problems.

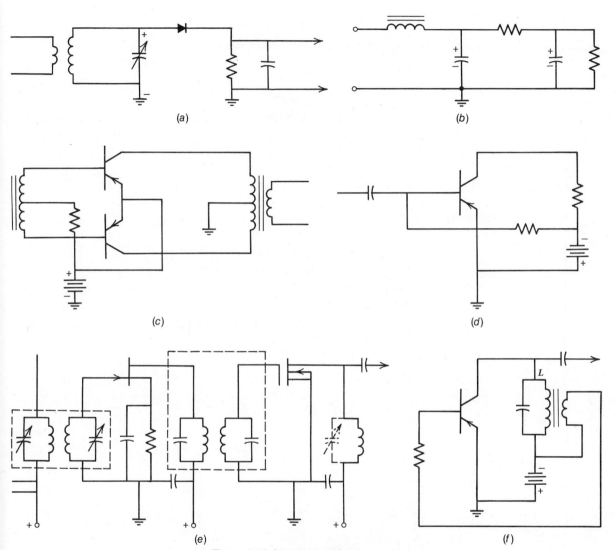

Figure 5-10 Circuit identification problems.

5-8 What is the purpose of the parallel coil and variable capacitor combination in the circuit in Figure 5-10a?

5-9 In the circuit (Fig. 5-10b), explain the function of the two capacitors.

5-10 (a) What is the polarity of the potential applied to the collectors of the transistors in the circuit in Figure 5-10c? (b) Explain your answer.

5-11 Why is a capacitor connected to the base of the transistor in Figure 5-10*d*?

5-12 Identify the type of interstage coupling used in Figure 5-10*e*.

Now return to the self-evaluation questions at the beginning of this chapter and see how well you can answer them. If you cannot answer certain questions, place a check next to them, and review the appropriate sections of the chapter to find the answers.

References **Bernard Grob,** *Basic Electronics,* Third Edition, McGraw-Hill, New York, 1971.

Milton S. Kiver, *Transistor and Integrated Electronics,* Fourth Edition, McGraw-Hill, New York, 1972.

Boyd Larson, *Transistor Fundamentals and Servicing,* Prentice-Hall, Englewood Cliffs, N.J., 1974.

H. Alex Romanowitz, and **Russell E. Puckett,** *Introduction To Electronics,* Wiley, New York, 1968.

Chapter 6

Drawing Schematic Diagrams

Instructional Objectives
1. To review electronic symbols.
2. To understand the need for conformity in drawing schematic drawings.
3. To present properly drawn schematic diagrams.
4. To develop an understanding of symmetry and balance in drawing a diagram.
5. To become effective in symbol placement and space arrangement for maximum legibility.
6. To draw the stages of a schematic diagram in proper sequential manner.
7. To develop an ability to convert a bread-boarded circuit into a proper schematic diagram.
8. To develop consistency in component code or acronym location on the schematic diagram.

Self-Evaluation Questions Test your prior knowledge of the information in this chapter by answering the following questions. Watch for the answers as you read the chapter. Your final evaluation of whether you understand the material is measured by your ability to answer these questions. When you have completed the chapter, return to this section and answer the questions again.

1. What is the purpose of a schematic diagram?
2. Why is a pictorial wiring diagram insufficient for understanding circuit functions?
3. How should a crossover be drawn when two conductor lines cross but do not connect?
4. What sequential layout of stages should be used?
5. What drafting practices and conventions are required in making schematic diagrams?
6. What is accomplished by observing principles of symmetry and balance in the location of components within a schematic diagram?
7. Waveforms are included in the schematic diagrams of certain

electronic circuits? Name these circuits.

8. What is the purpose of a *rotary switch function table?*

9. What drafting practices should be avoided in drawing proper schematic diagrams?

10. Give the purpose of reference designations and their location in a schematic diagram.

6-1
Cables, Junctions, and Crossovers

The schematic or elementary wiring diagram shows the functions and relations of circuit components through the use of graphical symbols. In circuit design, circuit analysis, and troubleshooting, it is more convenient to use schematics than pictorial drawings; this is especially true once the reader becomes familiar with component symbols and functions.

Since the purpose of a schematic diagram is to show the wiring of components in a circuit with utmost clarity and simplicity, the following drafting practices should be observed.

1. Most schematic diagrams are drawn using *medium-weight* lines. Occasionally, for instructional purposes, some lines or symbols, such as tube and transistor envelopes, are made heavier.

2. Parallel lines should not be drawn too near each other except when they are to be *cabled* as shown in Figure 4-9a.

3. Diagonal and curved lines are to be avoided. Horizontal and vertical lines present a more uniform and pleasing overall appearance than indiscriminate diagonal lines.

4. When a horizontal and a vertical line are intended to make an electrical connection, the *connection* is always indicated by a heavy dot *where the lines cross;* this simulates a soldered connection.

5. If it is intended that a horizontal and vertical *crossover* are *not* electrically connected, the dot should be omitted. (The older half-circle crossover is now considered both time-consuming and obsolete.)

6. When horizontal and vertical lines connect but do *not* cross over each other, the dot is not necessary.

7. A dashed or broken line is used to indicate a mechanical linkage or connection between components, such as variable capacitors or *decks* of *rotary switches* that operate on the same shaft.

8. Lines should be routed as directly as possible with a min-

imum of direction changes and crossovers.

9. The uppermost horizontal line of a schematic diagram should have the highest voltage relative to the lower bottom line or ground (see Section 6-2).

6-2 Grounds, Chassis, and Circuit Returns

Many of the components usually have a *common* connection which may be at the circuit's lowest potential or *ground*. It is advisable to observe the following procedures relative to common connections and circuit returns.

1. Very long lines or interconnecting leads are to be avoided; this may be accomplished through the use of chassis or earth grounds, as shown in Figure 4-10*b* and *c*.

2. When components are mounted on a metal panel or chassis, short common leads may be connected or soldered to the chassis; the chassis serves as the common lead.

3. Chassis and earth grounds should *not* be used interchangeably; the chassis ground sometimes may be at a potential higher (or lower) than the earth ground.

4. A *triangle* is the symbol used when the components are mounted on a nonconducting chassis and a common *bus* or wire is employed (see Fig. 6-1*a* and *c*).

6-3 Layout of Schematic Diagrams

Layout difficulties may occur when a schematic diagram contains a variety of different circuits with many interconnections. While some compromises may be necessary to avoid a confusing network of lines and crossovers, the following additional principles should be followed as closely as possible for *layout* of schematic diagrams.

1. Normal signal flow should be from left to right and from *top* to *bottom* (Fig. 6-1*c*).

2. Large multistage diagrams may be laid out in layers; signal flow should be from *left* to *right* (or top to bottom).

3. It is customary to begin with a *rough sketch* of the diagram observing normal signal flow (Fig. 6-1*b*).

4. Transistor and tube symbols should be aligned *horizontally*.

5. When several similar components (resistors, coils, capacitors, etc.) are connected to a common bus, they also should be aligned *horizontally* (Fig. 6-1*c*).

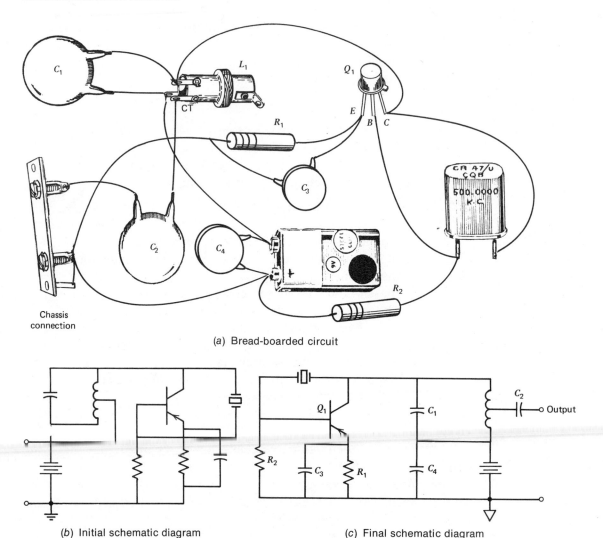

(a) Bread-boarded circuit

Chassis
connection

(b) Initial schematic diagram   (c) Final schematic diagram

Figure 6-1 Crystal oscillator circuits.

6. Achieve balance by *centering* a component within its *lead line* (Fig. 6-1c).

7. When components are connected in parallel, the *centers* of the component symbols should be aligned *horizontally*.

8. Quadrille or coordinate paper is often helpful in preparing a second rough sketch.

9. *Uniform density* of all graphic symbols is desirable; do not crowd symbols in one area while permitting large open

areas to exist on a drawing (Fig. 6-1*b*).

10. Check that all component connections are electrically accurate.

11. Letter and number all components to ensure that *all* are included in the final drawing in their proper location and connection in the schematic (Fig. 6-1*c*).

12. The final diagram must have a balanced, symmetrical, and pleasing appearance (Fig. 6-1*c*).

Refer to the crystal oscillator circuit of Figure 6-1; the breadboarded or pictorial wiring diagram (Fig. 6-1*a*) has been directly converted to the rough schematic diagram (Fig. 6-1*b*). Note that this initial diagram contains the following errors, which have been corrected in Figure 6-1*c*.

1. Output of the oscillator is at the left instead of at the right side of the diagram.

2. Transistor and capacitor symbols are incorrectly drawn.

3. Parallel lines are crowded.

4. Several symbols are not centered in their branch circuits.

5. There are unnecessary crossovers from resistor to bypass capacitor and from the crystal to the transistor base.

6. The top-center area of the diagram is crowded.

7. Improper chassis ground symbol.

8. The entire diagram lacks symmetry and balance; it is not pleasing in appearance.

Note that Figure 6-1*c* overcomes all these difficulties.

**6-4
Rotary
Switch
Layout**
Rotary switches may be in the form of a single *deck,* or an assembly of several decks connected by a single shaft; they find application in almost all electronic test equipment. Each switch position is determined by a mechanical device or *detent* placed at 30° intervals. The stationary decks contain the fixed contacts, while the deck *rotors* carry the movable contacts that connect the fixed contacts (Fig. 6-2).

Rotary switches, particularly those with more than one deck, are apt to present circuit reading problems unless considerable care is used in terminal identification.

1. Each contact terminal must be assigned a number.

2. The numbering sequence should preferably be in the direction of rotation.

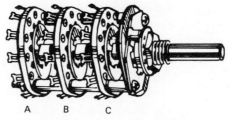

(a) Three deck rotary switch

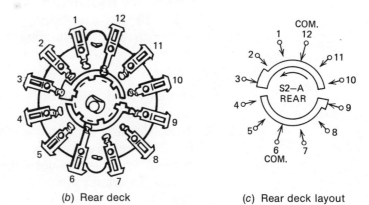

(b) Rear deck

(c) Rear deck layout

(d) Rear deck symbol

(e) Rear deck symbol

	S2—A	
Pos.	Range, MHz	Terminals
1	26—31	12—1, 6—7
2	31—37	12—2, 6—8
3	37—44	12—3, 6—9
4	44—54	12—4, 6—10
5	54—88	12—5, 6—11

Figure 6-2 Rotary switch layout and symbols.

(f) Function table

3. The direction of rotation must be indicated with an arrow (Fig. 6-2c, d, and e).

4. The deck viewpoint must be included such as *front, rear,* F, or R. The front designation would apply to the switch deck surface nearest the switch control knob.

5. All *in-line* terminals on a multiple deck rotary switch should have the same number; however, each deck should be assigned a letter. For example, a terminal may be designated as S2–B,6, meaning terminal 6 on switch S2 and deck B.

Ideally, multiple decks should be grouped together on the schematic diagram but not at the expense of long circuit lines. Since each deck is only mechanically attached to another deck of a multiple switch, it is permissible to separate each deck from the others on the same schematic diagram if each deck is properly labeled. Each deck may be joined by a broken line to signify a common mechanical linkage as shown in Fig. 6-2d and e.

Figure 6-2c is a semipictorial rotary switch symbol of a single deck; note the mechanical similarity to Figure 6-2b. Simpler forms of switch layout of the same deck are illustrated in Figure 6-2d and e; however, circuit reading or tracing is less difficult when the layout shown in Figure 6-2c is used.

The lower switch circuit (Fig. 6-2c) with a 6 common terminal indicates a rotary switch in which the contact connection is opened *before* connection is made with an adjacent contact; these are commonly known as *nonshorting* or *break-before-make* switches. Occasionally, it is desired that connection *is* made with the adjacent contact *before* the first connection is opened; in such a case, called *make-before-break,* the arrow of the rotating contact is replaced with a shorting bar as shown in the upper switch circuit with a common 12 terminal.

To further simplify circuit reading problems, the relationship between switch positions and circuit functions should be given on the drawing in tabular form as shown in Figure 6-2f. Note that the decks are all in position 3 of the *function table; switch blade 12* is contacting terminal 3, and blade 6 is contacting terminal 9. This position also indicates that the circuit is operating within the 37 to 44 MHz (megahertz) range.

6-5
Identification of
Components
Even the simplest schematic diagram is apt to contain more than one similar symbol. It is therefore important to identify every component on a schematic diagram to distinguish it from a similar component. Good practice dictates observation of the following procedures.

1. A code letter and number such as $R3$ may be used to identify the third resistor in a diagram. (See Fig. 6-1c and Appendix A for acceptable component codes.)
2. When codes are used, the code must be explained in a parts list.
3. A component may be identified by means of its specification such as 47 kΩ (kilohm), 1 watt for a resistor.
4. Adequate space must be available near each symbol for such identification.
5. Identifications must be consistently placed adjacent to the same side of each symbol (Fig. 6-1c).
6. When codes are used, each class of component should be numbered progressively from left to right across a drawing. This practice simplifies location of the component providing that the parts list is tabulated in numerical order (Fig. 6-1c).

Schematic drawings for troubleshooting complex circuits, such as in television receivers, usually include small *waveform* drawings at the important *test points*. The illustrated waveform is typical of that obtained when an *oscilloscope* is connected from the test point to ground.

Diagrams specifically designed for servicing also include typical voltages measured at each transistor terminal and other strategic test points. A *vacuum tube voltmeter* (VTVM), an *electronic voltmeter* (EVM), or a similar *nonloading* voltage-measuring instrument should be connected between the test terminal and ground to verify the referenced voltage.

Summary
1. Schematic diagrams show the interconnections of electronic symbols necessary to form a functioning circuit.
2. Observance of several drafting conventions is necessary to obtain an easily read schematic diagram.
3. Conventional treatment of these diagrams determines signal flow direction, arrangement, and spacing of components.
4. Codes or other forms of component identification are always included for maximum utilization of a schematic diagram.

5. Additional material, such as test voltages and waveforms, is often included in a schematic diagram.

Problems **6-1** List as many drafting errors as you can find in the improperly drawn schematic diagram of Figure 6-3.

6-2 Properly prepare a schematic diagram of Figure 6-3.

6-3 Prepare a rotary switch layout similar to Figure 6-2c in which each of two decks contain three circuits of four contacts each; a function table must be included.

6-4 through **6-10** These problems refer to Figures 6-4 through 6-10, respectively. Draw a proper schematic diagram of each pictorial diagram; include a parts list. The components are to be coded instead of identified by specifications.

Now return to the self-evaluation questions at the beginning of this chapter and see how well you answer them. If you cannot answer certain questions, place a check next to them, and review the appropriate sections of the chapter to find the answers.

References **Charles J. Baer,** *Electrical and Electronics Drawing,* Third Edition, McGraw-Hill, New York, 1973, pp. 129–158.

Nicholas M. Raskhodoff, *Electronic Drafting and Design,* Second Edition, Prentice-Hall, Englewood Cliffs, N.J., 1966, pp. 358–382.

George Shiers, *Electronic Drafting,* Prentice-Hall, Englewood Cliffs, N.J., 1962, pp. 261–338.

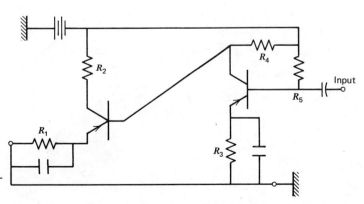

Figure 6-3 Improperly drawn direct-coupled amplifier.

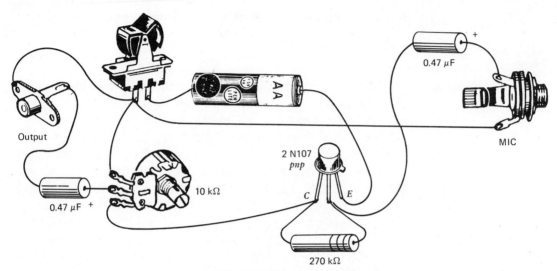

Output

0.47 µF +

10 kΩ

2 N107
pnp

C   E

270 kΩ

0.47 µF

+

MIC

Figure 6-4 Microphone preamplifier circuit.

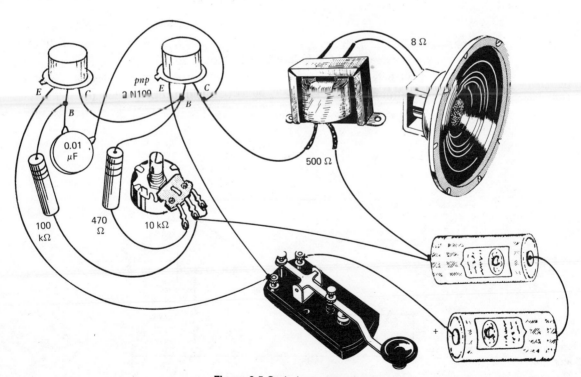

E   C

pnp
a N109   E   C

B   B

0.01
µF

8 Ω

500 Ω

100
kΩ

470
Ω

10 kΩ

+

Figure 6-5 Code instructor circuit.

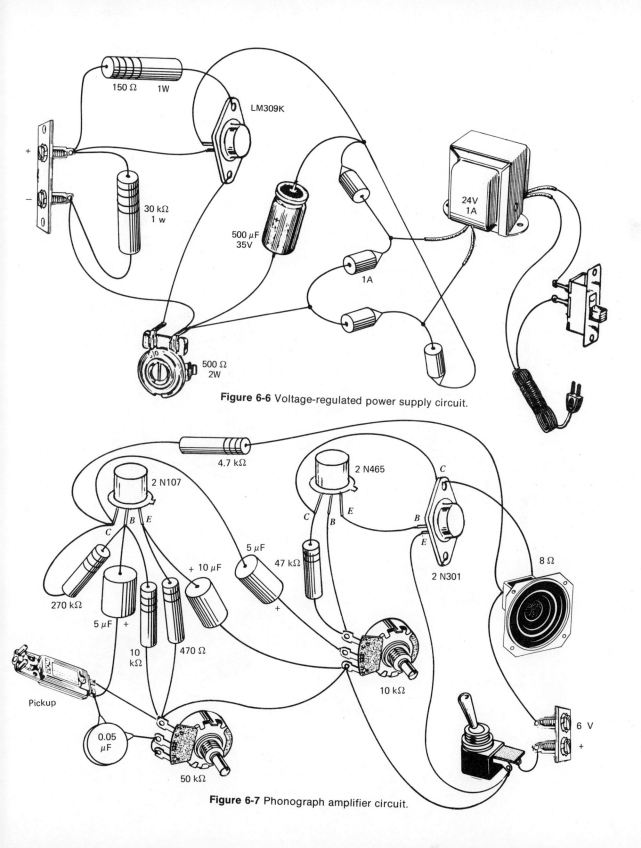

150 Ω 1W

LM309K

30 kΩ 1 w

+

−

500 μF 35V

24V 1A

1A

500 Ω 2W

Figure 6-6 Voltage-regulated power supply circuit.

4.7 kΩ

2 N107

2 N465

C

B E

C

C B E

B E

2 N301

8 Ω

270 kΩ

5 μF +

+ 10 μF

5 μF

47 kΩ

10 kΩ

470 Ω

10 kΩ

Pickup

0.05 μF

50 kΩ

6 V +

Figure 6-7 Phonograph amplifier circuit.

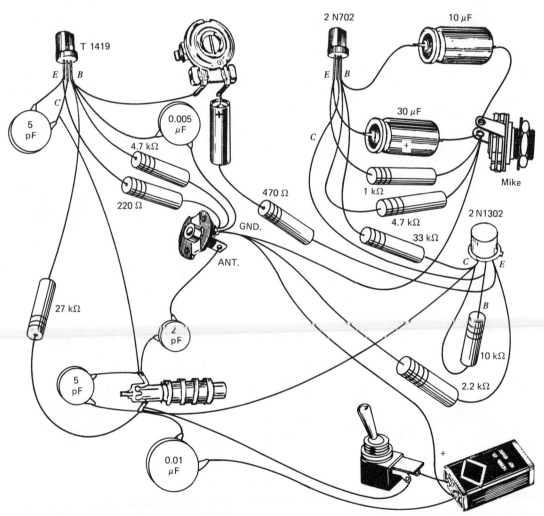

Figure 6-8 FM transmitter circuit.

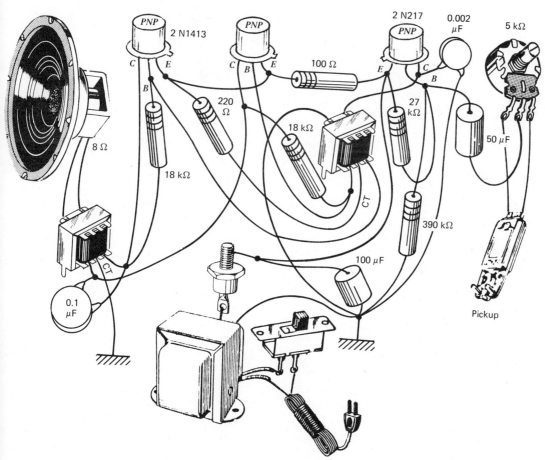

Figure 6-9 Push-pull type phonograph amplifier.

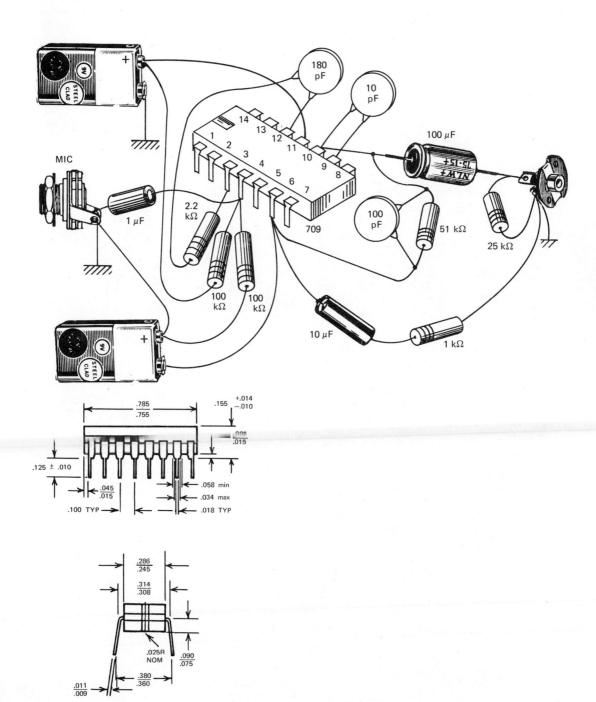

Figure 6-10 Operational amplifier integrated circuit.

Chapter 7 Wiring Assembly Diagrams

Instructional Objectives
1. To learn the advantages of cable and harness wiring.
2. To become familiar with the wire table.
3. To understand the relationship between current-carrying capacity and wire size.
4. To become proficient in planning and layout of wire assemblies.
5. To develop an understanding of wire-coding systems.
6. To learn how to prepare a wire-harness table.
7. To understand the difference between a baseline connection diagram and a highway connection diagram.

Self-Evaluation Questions Test your prior knowledge of the information in this chapter by answering the following questions. Watch for the answers as you read the chapter. Your final evaluation of whether you understand the material is measured by your ability to answer these questions. When you have completed the chapter, return to this section and answer the questions again.

1. What factors determine the use of wiring harnesses?
2. What is the difference between an airline connection diagram and a schematic diagram?
3. In which type of wiring assembly diagrams are we concerned with trunk and feeder lines?
4. What factors should be considered when choosing a wire size?
5. What is included in a wire code?
6. In what applications is stranded wire used?
7. What are the determining factors in choosing wire insulation?
8. What is accomplished through the use of coaxial cable?
9. What data should be included in a wire-harness table?
10. What is the purpose of cabling and how is it done?
11. How are wire-harnesses constructed?

7-1
Wires and
Cables

Solid, *tinned,* insulated copper wire is almost always used in wiring electronic devices containing large *discrete components.* Bare wire is often used for the very short connections between adjacent terminals and for long common connection wires. Instead of *stripping* the insulation from both ends of short wires, bare wire is often insulated with varnished cotton or rayon *sleeving,* commonly known as "spaghetti." Chapter 8 will discuss *printed circuits;* in this wiring method, flat, copper-plated conductors on a flat insulating material carry the current. Copper conductors are usually coated with tin or a tin-lead alloy to minimize corrosion and to ensure good solderability.

The size of round wire is specified by its American Wire Gauge (AWG) number. As noted in Appendix B, the smaller gauge numbers denote large diameter wires. Conversely, small diameter wires are identified by the higher gauges. For example, No. 20 *hookup* wire has a diameter of 0.80 mm (0.0315 in.), while No. 18 wire has a diameter of 1.0 mm (0.0394 in.).

Larger diameter wires carry more current and have less resistance than smaller diameter wires of the same length without dangerous heating. The recommended *current-carrying capacity* of copper wire is included in Appendix B. Note that No. 20 hookup wire may only carry 5 amperes safely, while No. 18 wire may be used with currents up to 7 amperes.

Stranded wire is specified when the conductor is subject to vibration or movement, since a solid wire may break in such an application. Stranded wire is fabricated by twisting many bare small diameter wires together; stranding thereby promotes flexibility. The equivalent solid gauge number of stranded wire is determined by obtaining the product of the strand's *cross-sectional* area or *circular mil* area and the number of strands. The circular mil area of round wire is obtained by squaring the *mil diameter* or diameter in thousandths of an inch.

Example 7-1
Find the solid copper AWG size wire equivalent to 7 strands of No. 32 AWG wire.

Solution
1. Using Appendix B, find the strand diameter in millimeters as AWG No. 32 = 0.02 cm = 0.2 mm.

2. Calculate the circular millimeters (c-mm) by squaring the strand diameter

$$c\text{-mm} = (0.2 \text{ mm})^2 = 0.04 \text{ c-mm}$$

3. Calculate the total circular-millimeter area by multiplying the strand diameter area (in circular millimeters) by the number of strands in the wire

$$A_{c\text{-mm}} = 7 \text{ strands} \times 0.04 \text{ c-mm/strand} = 0.28 \text{ c-mm}$$

4. Reconvert the total area to an equivalent diameter in millimeters by taking the square root of the circular-millimeter area, or

$$D = \sqrt{A_{c\text{-mm}}} = \sqrt{0.28 \text{ c-mm}} = 0.053 \text{ mm}$$

5. Using Appendix Table B find the AWG No. equivalent, or a diameter of

$$0.053 \text{ mm} = \text{No. 24 AWG (approximately)}$$

When conductors are carrying currents from different circuits, they must be insulated from each other to prevent *short circuits*. Insulation may consist of enamel, cotton, nylon, fiberglass, and many plastic coatings. The type, grade, and wall thickness of insulation is determined by electrical factors such as voltages, operating frequencies, *dielectric losses,* and environmental conditions.

7-2 Cable Drawings To prevent high-frequency signals from being induced in or radiated from a conductor, the conductor insulation is covered with a grounded, woven metallic braid; it is then known as a *coaxial cable.* This outer metallic sheath serves both as a high-frequency *shield* and as a common ground conductor. The symbol for coaxial cable is shown in Figure 4-9*b*; the grounded sheath is indicated by the grounded circle surrounding the conductor. Special care should be taken at the *terminations* of coaxial cable; male and female connectors, shown in Figure 4-9*c*, are always specified. Single conductor or coaxial cables are used in radio-frequency circuits for antenna lead-ins, transmission lines, and other intercircuit connections.

Several conductors may be cabled or enclosed in a *sheath,* in which case all the conductors are encircled in the schematic diagram. *Multiple* conductor cables require an *assembly drawing* as shown in Figure 7-1.

Connectors for multiple conductor cables are available in many styles and sizes. A typical *male* connector is shown in Figure 7-1*a*; the mating *female* connector contains matching holes for the *male pins* or inserts. Figure 7-1*b* shows a female connector designed to be *panel mounted*. These connectors are rigidly coupled by means of a *locking ring* and pin. The cable is rigidly fastened to each connector with a threaded squeeze fitting; some connectors use cable clamps. Note that the parts list includes the *pin location* of each different colored wire.

With the advent of IC modules, manufacturers have taken advantage of the availability of 14 and 16 pin IC sockets in an attempt to standardize and reduce the cost of cable connectors. The flat, molded, multiple conductor cable shown in Figure

(a) Male connector

(b) Female panel connector

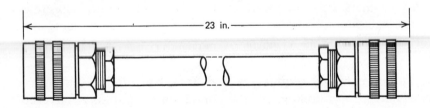

(c) Cable drawing and parts list

Req D	Part No.	Description	Length (in inches)	Left Connector Pin	Right Connector Pin
2	MS 1234	Male connector			
1	DWG 654	Plastic connector	21		
1	DWG 655	White wire	22	A	E
1	DWG 656	Red wire	22	B	D
1	DWG 657	Blue wire	22	C	C
1	DWG 658	Green wire	22	D	B
1	DWG 659	Black wire	22	E	A

Figure 7-1 Cable details.

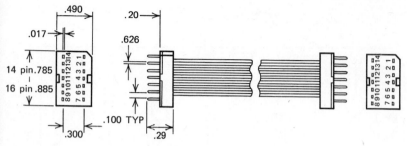

(*d*) Integrated circuit interconnect cable

Figure 7-1 (*Continued*)

7-1*d* has the advantage of less bulk and may be placed beneath a chassis. These *ribbon cable assemblies* are available with color-coded 26 AWG stranded wire.

**7-3
Point-to-point
Connection
Diagrams**

In this simple method of wiring, individual color-coded wires are strung directly between component terminals. It is a convenient system for simple assemblies and has the advantage of shorter and more accessible leads. Undesirable interaction of circuits is avoided since only a few wires are strung parallel or close to each other.

As shown in Figure 7-2*a,* many of the nonessential details of the components are omitted in the drawing; however, the *size* and *location* of components must be *drawn to scale;* such a procedure permits *direct* measurement of *wire lengths.*

The wires of all types of connection diagrams must be coded for identification and simplification of assembly. The simple coding in Figure 7-2*a* has a *space* advantage. However, it has the disadvantage of requiring an expanded wiring list, as is partially shown in Figure 7-2*b*. The wiring list must contain all the details concerning each wire as shown in Figure 7-2*b*, but the origin and destination columns need not be included because such information may be obtained from the diagram.

The wire color may be abbreviated, or it may be identified using the resistor color code or the numerical code recommended by Mil Std 122:

Black	BK	0		Green	GN	5
Brown	BR	1		Blue	BL	6
Red	R	2		Violet	V	7
Orange	O	3		Gray	GY	8
Yellow	Y	4		White	W	9

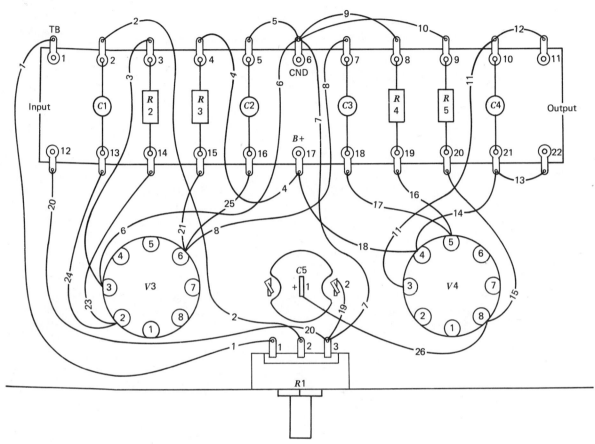

(a) Point-to-point wiring drawing

Wire Code	Color	Length, cm	Size AWG	From Component	Term.	To Component	Term.
1	BK	14	22	TB	1	R1	1
2	BL	11	22	TB	2	R1	2
3	BR	7	22	TB	3	V3	3
4	BK–R	7	22	TB	4	TB	17
5	BARE	2	22	TB	5	TB	6

(b) Partial point-to-point wiring list

Figure 7-2 Point-to-point wiring.

Many assemblies require more than ten different wires; it is therefore necessary to expand the wire color-code list through the use of color combinations. A *tracer* thread of a contrasting color is added to the outer insulating braid, thus vastly increasing the available number of different coded wires. The tracer color code is placed after the body color code, as shown in Figure 7-2*b*, item 4.

The length of each wire includes the amount of insulation that has been stripped in preparation for soldering to a terminal. A note should be added specifying the amount of stripping. It may also be necessary to specify the location and type of a few critical wires that must withstand high voltage or may cause circuit interactions. When a wire is placed at about 90° to the other wires, radiation and induction is kept to a minimum.

**7-4
Baseline
Connection
Diagrams** Point-to-point layouts become impractical when many wires must be shown within a compact assembly. Many military, industrial, and commercial specifications do not permit free and random wiring; all wiring should be properly secured and completely identified. *Cable harness* or *cableform* wiring satisfies these requirements in that many wires of different types, colors, and lengths are grouped and laced together into a fairly rigid cable. Such wiring methods increase circuit reliability and facilitate maintenance. Although additional labor costs are involved in producing a cable harness, less time is needed for assembly and connection of the cable harness than for loose wires. When specifications are not rigid, small-volume manufacturers avoid the additional cost of cable harnesses since their assemblers may be more capable of reading a wiring assembly diagram.

The *baseline,* or *airline,* wiring diagram usually contains a single horizontal or vertical cable conveniently located so that the short *feed lines* may be drawn from the cable to all component terminals. The same circuit that was used in the point-to-point diagram (Fig. 7-2) has been redrawn in Figure 7-3 in baseline form.

Note that more wire is needed in Figure 7-3, the baseline diagram, than in the point-to-point wiring of Figure 7-2. An attempt at economy has been made by using short *jumper wires,* as between terminals 13 and 14, to eliminate one long wire (No.

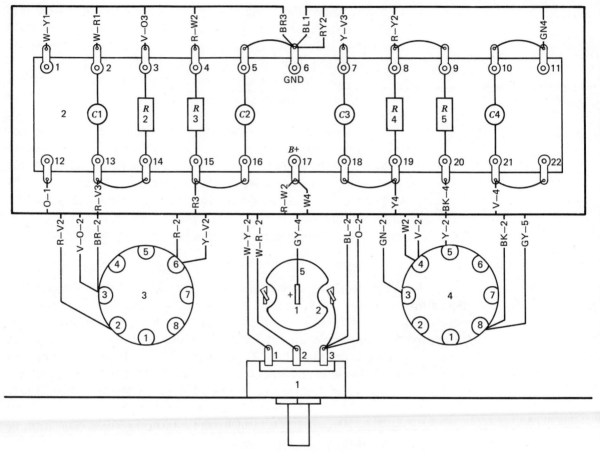

Figure 7-3 Airline, or baseline, wiring diagram.

23) from the same terminal 2 of *V*3. Furthermore, the left- and right-hand cable loops might be avoided if the wires to terminals 1 through 11 of the terminal board (2) could travel directly upwards between the components. This last method is often found undesirable since it may cause service problems. For purposes of illustration, the highway diagram shown in Figure 7-4 was drawn using direct feed lines to the TB terminal board (see Section 7-5).

The wire code used in Figure 7-3 is a color and destination code. For example, the GY-5 wire attached to terminal 8 of component 4 indicates that a gray wire should be used and connected to component 5. The code does not specify which terminal of component 5 is to be connected to the gray wire. Beginning instead at terminal 1 of component 5, note that the code is GY-4; thus, it specifies that the gray wire should be attached to some terminal of component 4. Since the wire code does not include the complete destination description, wire lengths, and wire sizes, a wiring list or *routing chart* must accompany the baseline wiring diagram.

7-5
Highway
Connection
Diagrams
The *highway* type connection diagram may eliminate some assembler errors since all the wires to a component are cabled into a smaller cable highway. Furthermore, the highway diagram (Fig. 7-4) uses a more complete wire coding than the previous diagrams.

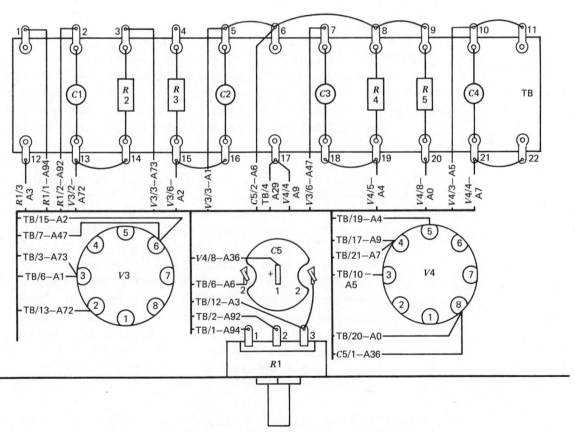

Figure 7-4 Highway-type connection diagram.

The ANSI-recommended wire code, such as TB/13-A72, requires that each feed line contain the following information near the point where it joins the trunkline or highway.

1. Component destination; terminal board, TB.
2. Terminal destination; terminal 13 of TB.
3. A slanting line should separate the component destination and terminal: TB/13.
4. Wire size code; A may mean 20 AWG.
5. Numerical wire color code; 72 indicates a violet wire with a red tracer.

The length of each wire is *not* included in the code; it is therefore necessary to include a wiring list in the highway drawing. This wire code has the obvious disadvantage of requiring more lettering space within the feeder lines.

**7-6
Cable
Harness
Construction
Drawings**
A *jig* or board containing pins at the feed line *branch-off points* must be constructed for the fabrication of identical cable harnesses. A simple full-scale orthographic drawing of the jig construction is usually required. Dimensions to centerlines of cables and feeders and the lengths of cables and feeders must be included and referenced from the horizontal and vertical cables.

When the ends of feeder lines are to be stripped, the amount of insulation removed must be specified. Some manufacturers do not solder the feed lines to the component terminals; another method requires the *terminal lugs* to be soldered or mechanically attached to the feed lines before assembly of the harness into the chassis. The drawing and parts list must include the terminal lug details and specifications.

Before the wires are removed from the cable harness jig, it is necessary to secure all cabled wires together. For small production quantities, waxed twine or nylon cord is laced and tied about the cabled wires. Plastic ties, preformed sleeving, and spiral plastic bindings are also employed when the additional expense is warranted by large production runs.

Summary 1. Wiring assembly diagrams are production drawings required for the purchasing, assembly, and maintenance of connecting wires and terminations.

2. Point-to-point wiring uses the least amount of wire but requires more assembly time and better-trained technicians than other methods of wiring.
3. Many wires of different types, colors, and lengths are grouped and laced together to form a cable harness.
4. After cable harnesses have been prepared, final assembly time of an electronic device is reduced.
5. Careful identification of all conductors must be included in all types of wiring assembly diagrams.
6. Wiring or routing lists include origin, destination, color, length, and size of all wires are included on wiring assembly diagrams.

Now return to the self-evaluation questions at the beginning of this chapter and see how well you can answer them. If you cannot answer certain questions, place a check next to them, and review the appropriate sections of the chapter to find the answers.

Problems 7-1 How much current can a 0.200 cm diameter copper wire carry without overheating?

7-2 How much current can a 0.80 mm diameter copper wire carry without overheating?

7-3 What diameter copper wire is necessary to carry 7 amperes without overheating?

7-4 What solid copper AWG size is equivalent to 11 strands of 24 AWG copper wire?

7-5 How many strands of No. 38 AWG copper wire are needed to obtain the equivalent cross-sectional area of No. 16 AWG copper wire?

7-6 Complete the partial point-to-point wiring list of Figure 7-2*b*.

7-7 Copy the back panel view of the in-circuit transistor tester of Figure 7-5 on tracing paper. The overall dimensions should be 12 cm × 16 cm; draw and position the components to the same scale. Make several prints of the tracing; use one of them to add the point-to-point wiring of the tester. Prepare a complete wiring list.

7-8 Using another of the prints produced in Problem 7-7, add the baseline wiring of the transistor tester and wiring list.

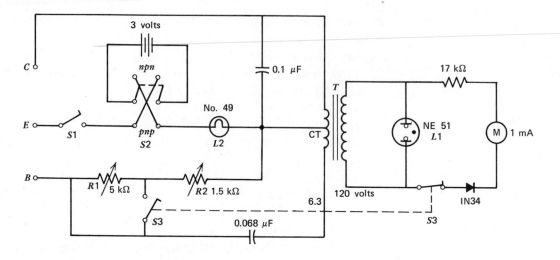

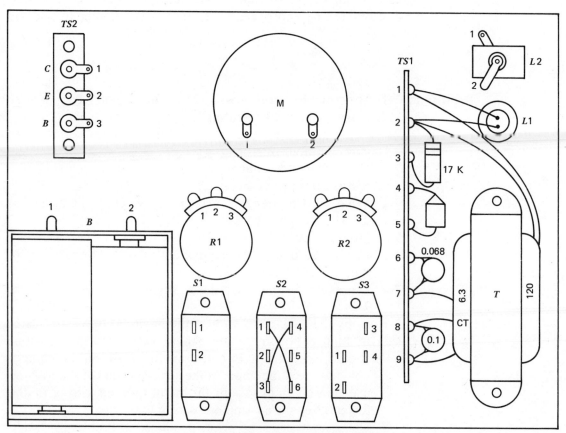

Figure 7-5 In-circuit transistor tester.

7-9 Add highway and wiring list of the transistor tester to still another print or on the original tracing.

7-10 Make a cable assembly drawing including parts list similar to Figure 7-1c of a 50 cm 58/U coaxial cable with a UG-88/U BNC plug and a UG-89/U BNC jack. The 58/U cable has a single solid 20 AWG shielded conductor and a 0.50 cm outside diameter. Refer to Figure 4-9 for an illustration of BNC connectors.

7-11 Make a cable assembly drawing including parts list similar to Figure 7-1c of a 40 cm braid shielded audio cable with a phono-plug similar to Figure 4-9g at each end. The cable contains two No. 20 ($^{10}/_{30}$ stranding) tinned copper conductors, colorcoded PVC insulation, tinned copper braid shield, and gray PVC jacket overall. Include a parts list.

7-12 Copy the Fig. 7-6 layout of a phonograph amplifier on tracing paper. The overall dimensions should be 11 cm $\times$ 18 cm; draw and position the components to the same scale. Make several prints of the tracing; use one of them to add the point-to-point wiring of the amplifier. Prepare a complete wiring list.

7-13 Add highway and wiring list of the Fig. 7-6 amplifier to still another print of the tracing drawn in Problem 7-12.

7-14 Repeat Problem 7-13 for baseline wiring and list.

References Charles J. Baer, *Electrical and Electronics Drawing,* Third Edition, McGraw-Hill, New York, 1973, pp. 69–91.

Nicholas M. Raskhodoff, *Electronic Drafting and Design,* Second Edition, Prentice-Hall, Englewood Cliffs, N.J., 1966, pp. 228–254, 446–478.

George Shiers, *Electronic Drafting,* Prentice-Hall, Englewood Cliffs, N.J., 1962, pp. 386–397, 421–457.

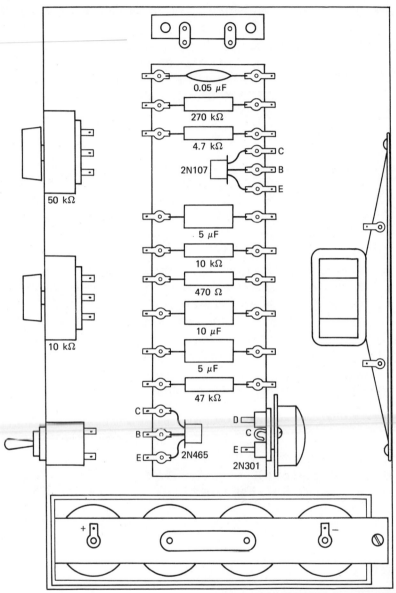

Figure 7-6 Phonograph amplifier component layout for Figure 6-7 circuit; to be used in Problems 7-12 to 7-14.

Chapter 8

**Printed
Circuit
(PC)
Drawings**

Instructional 1. To understand the need for miniaturization and printed circuits
Objectives (PCs).
2. To learn of advantageous applications of PC wiring.
3. To develop familiarity with printed circuit board (PCB) materials
and specifications.
4. To learn about PC design factors.
5. To develop some proficiency in the design of PCs.
6. To learn how to avoid PC crossovers.
7. To learn some disadvantages of PCs compared to discrete wiring.
8. To learn construction methods of prototype PCs.
9. To become aware of several production methods for construction of
PCs.

Self-Evaluation Test your prior knowledge of the information in this chapter by
Questions answering the following questions. Watch for the answers as
you read the chapter. Your final evaluation of whether you understand the material is measured by your ability to answer
these questions. When you have completed the chapter, return
to this section and answer the questions again.

1. Why is the width and thickness of a printed circuit conductor important?
2. What is the first and most important step in converting a schematic drawing to a printed circuit?
3. What is the purpose of *resist*?
4. What are *laminates*?
5. Discuss three different methods of applying acid-resistive lines
and symbols to a blank copper-coated board.
6. Why is the spacing between printed circuit board (PCB) conductors important?
7. What is gained when the PC conductors are permitted to spread
over unoccupied areas?
8. Why should sharp corners in a PC conductor be avoided?
9. What is the purpose of PC registration marks?

10. How can a PC assist in removing excess heat?

**8-1
Discrete
Versus
Printed
Circuit
Wiring**

A printed circuit (PC) is a prefabricated conductive circuit pattern reproduced on a thin sheet of insulating material. The circuit pattern may be *offset printed* upon the copper-clad *laminated* sheet, but many other application methods are also used. PC wiring has the following advantages over point-to-point or harness wiring.

1. Simpler and less costly miniaturization.
2. Lower cost of wiring for large-volume production.
3. Lower assembly cost when automatic component insertion and soldering is used.
4. Reliability is improved because conductors remain in the same position.
5. No conductor interconnection errors.
6. Lower weight because of miniaturization, less hardware, and less wire insulation.
7. Components are easily coded and identified.
8. The wiring is more easily protected against the environment.

On the other hand, discrete circuit wiring has advantages over PC wiring in many applications; the following disadvantages of PC wiring cannot be overlooked.

1. Layout and design are more difficult.
2. Small quantity PCBs are more expensive to produce.
3. Circuit design of PCs is usually restricted to one plane.
4. More labor costs are involved for PC changes and repairs.
5. Subcontracting of PCBs is more difficult and costly.

**8-2
Materials and
Processes**

Printed circuit boards (PCBs) usually consist of laminated plastic sheets to which a thin sheet of copper is *bonded* on one or both sides. A common specification for the laminated material is NEMA grade XXXP paper base phenolic, clad with *1 ounce per square foot* of copper foil. This weight of copper corresponds to an actual thickness of 0.034 mm (0.00135 in.). Two-ounce cladding has a foil thickness of 0.068 mm (0.0027 in.).

Phenolic sheet thicknesses are obtainable from 0.80 mm ($^1/_{32}$ in.) to 6.35 mm ($^1/_4$ in.). A G-10 *glass epoxy* laminate is of better quality because the material has less tendency to warp from heat. Conductors may develop cracks or open circuits when-ever the circuit board warps. Glass epoxy board is *semitrans-parent;* this quality permits the components mounted on the unclad side to be visible from the copper-clad side.

If a single PCB is needed for a *prototype* electronic as-sembly, the conductor pattern may be drawn on the copper-clad surface with a felt-tip pen containing the *resist* (acid-resisting) fluid. Pressure-sensitive strip lines, circles, and pads (see Fig. 8-2) of acid-resisting material are obtainable and may be applied as in Figure 8-1.

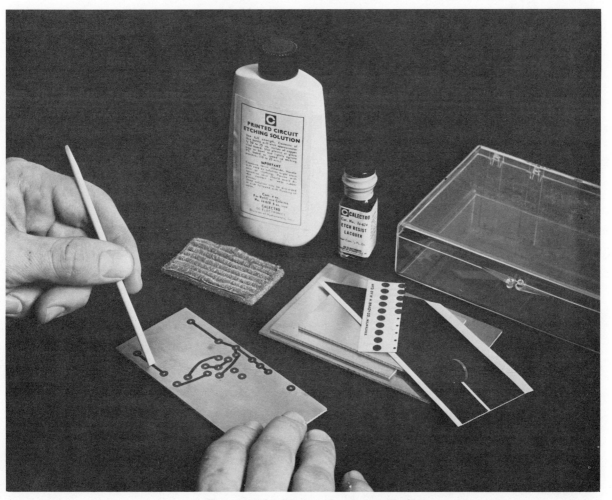

Figure 8-1 Application of resist appliqués to copper-clad boards. (Reproduced from *Popular Science* with permission © 1971 Popular Science Publishing Co.)

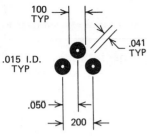

(a) 3-lead TO-5 pattern

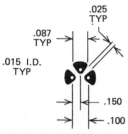

(b) 3-lead TO-16 pattern

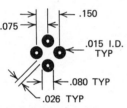

(c) 4-lead TO-5 pattern

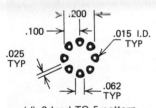

(d) 8-lead TO-5 pattern

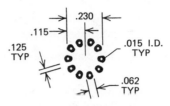

(e) 10-lead TO-5 pattern

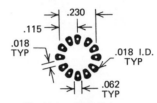

(f) 12-lead TO-5 pattern

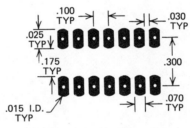

(g) 14-lead dual-in-line pattern

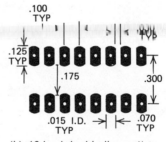

(h) 16-lead dual-in-line pattern

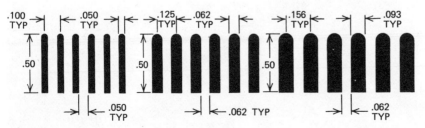

(i) Connector contact patterns

Figure 8-2 Typical resist appliqués.

The construction of a PCB occurs in the following steps.

1. Copper-clad surface is cleaned.
2. Conductor pattern is applied to the copper surface with an acid-resisting material or appliqué.
3. The uncoated copper areas of the PCB are removed with an etching solution (such as ferric chloride) as shown in Figure 8-3. The solution must be agitated (etchant molecules combine with the copper and become inactive) to bring fresh etchant in contact with the copper.
4. The completed etched board is cleaned to remove resist (using abrasion or a solvent) and expose the copper conductors.
5. The conductor pattern may be electroplated with gold, silver, or tin to facilitate soldering of components.
6. Holes slightly larger than component leads are drilled or punched at the conductor terminals or *lands* (Fig. 8-4). Larger holes may be drilled in at least each corner to mount the board and prevent warping. Some boards are edge-inserted and held by spring contacts; see Appendix I.
7. Electronic components are mounted on the unclad side with

Figure 8-3 Etching a printed circuit board. (Reproduced from *Popular Science* with permission © 1971 Popular Science Publishing Co.)

Figure 8-4 Drilling holes for component leads. (Reproduced from *Popular Science* with permission © 1971 Popular Science Publishing Co.)

their leads protruding through the terminal holes when using a single-clad PCB.

8. Component leads are trimmed, clinched, and soldered to the copper lands. A 25–30 watt soldering iron and 60/40 rosin-cored solder is recommended. When large quantities of PCBs are to be soldered, connections are simultaneously soldered by conveying and dipping the clad side in a flux and solder bath.

9. The copper-clad side of the PCB may be coated with an insulating lacquer.

If a small production quantity of circuit boards is involved, time is saved if photosensitive copper-clad boards are used. The copper surface of such boards are precoated with a *light-sensitive material*. When the board is exposed to light through a properly prepared *negative,* the *transparent* areas admit light

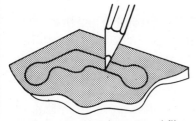

(a) Draw circuit pattern on red film

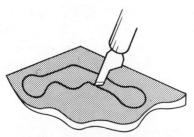

(b) Cut circuit layout through red coating

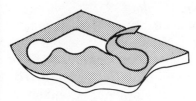

(c) Strip away red coating from conductor pattern

Figure 8-6 Preparation of red-film negatives.

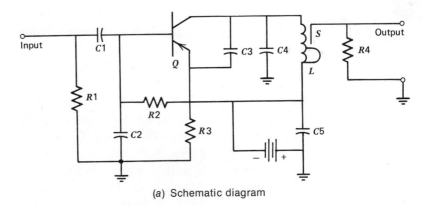

(a) Schematic diagram

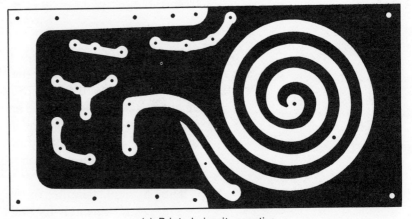

(b) Printed circuit, etched copper view, positive

(c) Printed circuit, negative

Figure 8-5 Impedance converter.

to the sensitive coating and cause it to become insoluble and harden in a developing solution. The insoluble portions form a protective barrier or resist against the action of the etching solution. Therefore, the clear areas of the *negative* (Fig. 8-5c) correspond to the conductor areas of the PC board. In effect, this is *reverse* printing.

A properly prepared negative permits *many* identical PC boards to be fabricated. If a photo laboratory is available to make photographic negatives and *reduce* the size of the layout, then this process is fast and desirable. A drawing of the circuit layout is made several times full scale on white paper. The white areas should represent the conductive areas; the drawing is then a negative of the conductor pattern. Finally, the drawing is reduced photographically to the desired size of the finished circuit board in the form of a *negative*.

A *positive* photosensitive *emulsion*-coated board is also available; its action is the reverse of the negative emulsion. The exposed areas are *not* hardened in the developer and are washed clear of emulsion. In this case, the drawing should be a *positive;* the conductor pattern will be *black*.

If photographic facilities are not available, a mechanical negative may be made by drawing the circuit layout on red film-coated plastic sheet. The drawing must be made full scale; for this reason, more care must be taken in its preparation. The red, coated sheet is placed over the scale drawing. Conductors and conducting areas are then outlined with a sharp (Fig. 8-6b) knife. The red "conducting" strips are then carefully peeled from the clear plastic sheet permitting light transmission.

8-3
Silk-Screen
Printing

Larger production quantities warrant the use of the *silk-screen* method for application of the resist. Top-quality silk is tautly attached to a wood frame. Silk is available in coarse (6XX) to very fine (20XX) weaves. The number of the silk refers to the mesh count or threads per square inch; a mesh count of 124 threads/in.^{2} (12XX) is recommended for printed circuits. The sizing must be removed from the silk by careful washing with a detergent.

The physical structure of the silk allows resist to be forced or "*squeegeed*" through its warp and woof. The *squeegee* is a simple tool containing a rubber or plastic blade ($^1/_4$–$^3/_8$ in.

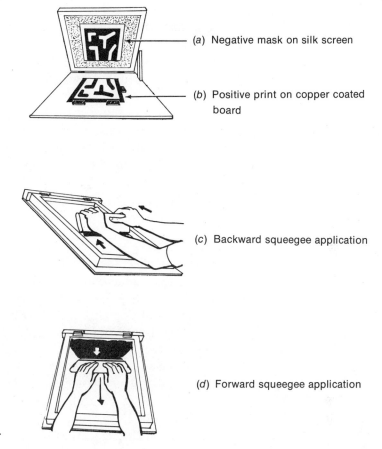

(a) Negative mask on silk screen

(b) Positive print on copper coated board

(c) Backward squeegee application

(d) Forward squeegee application

Figure 8-7 Silk-screen resist printing.

thick) inserted in a wooden handle. Areas *not* to be printed are *blocked out* with a resist paint on the silk screen itself. Thus the pores of the silk are closed wherever desired, preventing the resist from penetrating at those points (Fig. 8-7). If a copper-clad board is placed underneath such a silk-screen stencil before the resist is forced through the mesh, the resist will be deposited on the copper-clad board when squeegeed.

When the circuit contains much detail, it can be applied photographically in the following manner.

1. A photographic *positive* and a light-sensitive gelatin film are locked together in a photo-exposure frame, emulsion sides facing each other.
2. Exposure to light *hardens* all the gelatin areas under the

transparent parts of the photographic positive and leaves undisturbed those sections under the *opaque* areas of the positive.

3. The gelatin film is removed and *developed* or *etched* in hot water; the unhardened areas of the gelatin is washed out. The light-hardened areas of the gelatin act as a resist, thus becoming a photographic negative.
4. When the gelatin is dry, it is transferred to the underside of the mounted silk screen.
5. After firm contact is made with the silk, the backing is removed from the gelatin stencil.
6. PCBs are then printed in the usual manner.

8-4 Design Factors In Section 8-2, the weight and thickness of PC conductors was discussed. The conductors on a PCB are approximately rectangular in cross section instead of circular. Therefore, the thickness of copper foil and width of the conductor determines its current-carrying capacity. The thickness of copper foil has been conventionally specified as the weight of copper in ounces per square foot as listed in Table 8-1. Note that one of the conventionally used foil weights of 33.5 mg/cm² (1.0 oz/ft²) has a foil thickness of 0.0343 mm (0.00135 in.),

Since the line width also determines the current-carrying capacity, Table 8-2 gives the capacity for the two foil weights most commonly used.

It is recommended that the conductor paths are drawn as wide as possible as an economy measure; when a considerable amount of copper must be etched away, more etching solution is consumed, and the etching time is increased. Furthermore, large areas of copper will serve as *heat sinks* to remove and radiate heat from the components.

On the other hand, the spacing *between adjacent conductors* should never be less than 0.80 mm ($^1/_{32}$ in.) to prevent arcing between conductors. This minimum conductor spacing will permit a *peak* potential difference up to 150 volts. Doubling the spacing will approximately double the permissible potential difference between conductors.

Sharp corners or right-angled bends, as illustrated in Figure 8-8*a* and *b*, should be avoided to prevent *expansion cracks* in

Table 8-1
Thickness and Weight of Copper Foils for Copper-Clad Laminates

Foil Thickness		Foil Weight	
mm	in.	mg/cm^2	oz/ft^2
0.0173	0.00068	16.74	0.5
0.0343	0.00135	33.5	1.0
0.0686	0.0027	66.95	2.0
0.107	0.0042	100.4	3.0
0.173	0.0068	167.4	5.0
0.239	0.0094	234.3	7.0

Table 8-2
Current-Carrying Capacities for Two Common Copper-Clad Laminates

Line Width		Maximum Current (amperes)	
mm	in.	1 oz/ft^2	2 oz/ft^2
6.35	0.25	5.8	6.5
3.175	0.125	4.5	5.0
1.588	0.0625	2.5	3.5
0.794	0.0313	1.7	2.5

the conductors. As the temperature rises, the copper conductors tend to lengthen more than the laminated board, resulting in cracks in the conductor or destruction of the bond between the copper and the laminated board.

Terminal pads or *lands* are used for the connection of the PC to other circuit boards, components, or sources of power. Computer and TV circuit boards are often edge-connected through pressure contacts, as shown in Figure 8-2*i* and Appendix I. The pads should be wider than the conductor paths for convenience in soldering and better pressure contact. Occasionally, some terminal pads are reinforced with brass eyelets (see Appendix H) or terminal studs. Except for ground pads, all other pads should be displaced at least 0.80 mm ($^1/_{32}$ in.) from the edge of the board. This procedure prevents short-circuiting of the pads to the side of a metal case or adjacent components.

Round pads, lands, or *donuts* are placed for internal connection of board-mounted components. Usually these lands have a diameter of at least 2.4 mm ($^3/_{32}$ in.) except for the connections to integrated circuit (IC) modules (Fig. 8-2). Since the standard pin spacing for IC modules is 2.5 mm (0.100 in.), these pads are usually limited to a 1.5 mm ($^1/_{16}$ in.) width.

Many commercial PCs contain nonsymmetrical conductor patterns; it is more important that the pattern is designed to avoid the crossing of circuit paths. While crossovers are undesirable, it is possible to make a crossover using insulated wire (Fig. 8-5*b*). In Figure 8-8*f* and *h*, the insulated resistors act as crossovers when the conductor path is placed beneath the resistors. Crossovers may also be arranged if copper foil has been bonded to both sides of the laminate. Connection is made to the bottom conductor through brass eyelets (see Appendix H).

**8-5
Circuit
Board
Layout**

The first step in making a drawing for a PC is to study the schematic diagram. If you are not familiar with the circuit, obtain advice regarding compatibility of various sections of the circuit. Some components should not be placed near each other to prevent magnetic or capacitive coupling. High potential differences between certain conductors will make it necessary to increase the space between such conductors. Check the amount of current carried by each conductor to determine the necessary conductor width and thickness.

After component compatibility has been determined, the PCB layout procedure is as follows.

1. From the schematic diagram, determine which components are directly connected to a common ground.
2. The pattern layout is always drawn on the copper-clad side; this will mean that you are looking at the bottom of your components, that is, the *underside* of the PCB.
3. Using grid or quadrille paper, sketch these components to scale and position them near a common ground strip at the edge of the board; make an attempt at achieving symmetry (Figs. 8-5 and 8-11).
4. Check if several components are directly connected to the highest potential line; such a possibility may result in the need of another common potential strip. This strip may be

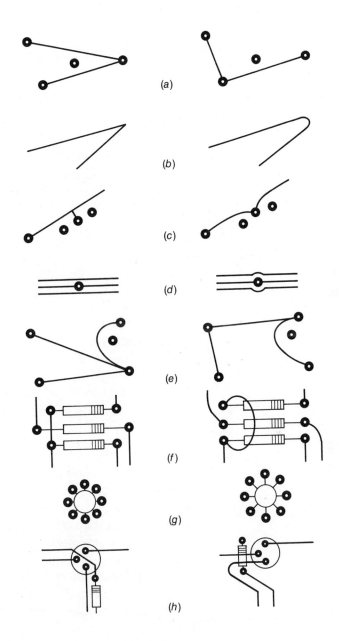

(a)

(b)

(c)

(d)

(e)

(f)

(g)

(h)

Figure 8-8 Improper and recommended layout details.

positioned at the edge of the board opposite the common ground.

5. Locate and sketch components, other than resistors or capacitors, which have point-to-point connections and do not break with a resistor or capacitor. It is important to locate these connections at an early stage of sketching because they probably cannot be crossed or jumped with a resistor.

6. Locate board terminations and power pads at an edge of the board. The location, size, and spacing of these pads may be dictated by the type of PCB pressure connectors used (see Fig. 8-2).

7. Make the remaining connections. If any conductor positions result in a conductor crossover, try to reroute the conductor under an insulated resistor or capacitor.

8. Recheck the PC layout against the schematic diagram to be sure that it is electrically correct.

9. Slightly change the position of conductors to agree with the recommendations shown in Figure 8-8.

10. Replace sharp bends with smooth, wide-curved conductors.

11. Increase the width of conductors located in large open spaces, especially those that are connected to transistor collector terminals.

The complex PCB shown in Figure 8-9 summarizes the conversion steps from schematic diagram to printed circuit. Particularly note the crossover elimination by crossing beneath components $C3$, $C4$, $D5$, $R2$, $R3$, $R4$, and $R9$. The most common connection in the schematic is the top line from the transformer secondary. It was found more convenient to arrange this common, snakelike conductor throughout the PC.

The PC of Figure 8-9 includes + *reference marks*. Reference marks are used for correct location of the board in the drilling fixture or when component codes are printed at each component location. When it is necessary to use a laminate copperclad on both sides, registration reference marks are needed for perfect alignment of the upper and lower printed circuits.

The impedance converter PC shown in Figure 8-5 illustrates a method of using a printed conductor as a spiral-shaped (L) tapped inductor. Final adjustment of the circuit is made by changing the position of the insulated crossover tap for best

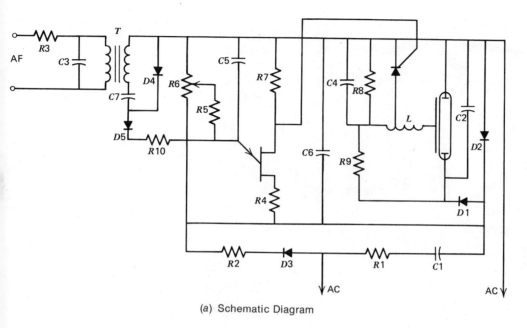

(a) Schematic Diagram

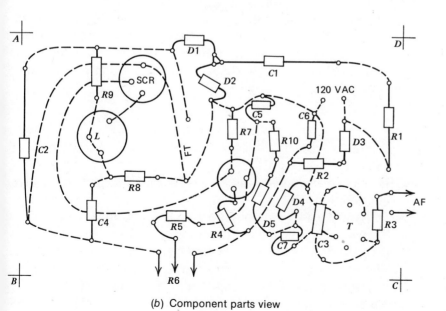

(b) Component parts view

Figure 8-9 Stroboscope diagram and PC layout.

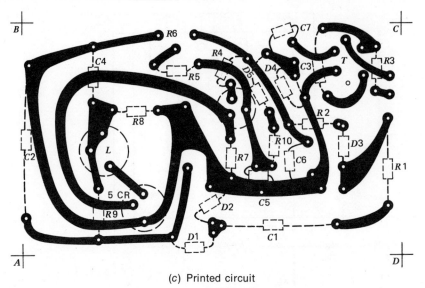

(c) Printed circuit

Figure 8-9 (Continued)

operation. Output from the circuit is obtained by induction through the shield conductor S.

The Heath Company printed circuit shown in Figure 8-10 illustrates a method of combining large components (such as rotary switches and potentiometers) on the PCB; many long, flexible leads are eliminated. This circuit board is fastened directly behind and parallel to the instrument panel. Also note that the PCB contains the shape and code number of each component.

Summary
1. Printed circuit (PC) wiring is used in the miniaturization of solid-state electronic circuits.
2. PCs lower the cost of interconnections for large volume production.
3. PCBs are ideal for component coding and location.
4. PC wiring is less subject to wiring and assembly errors than point-to-point or harness wiring.
5. The circuit conductors are drawn or printed with an acid-resisting ink on copper-clad, laminated plastic sheets.
6. An etchant solution removes all the copper cladding that has not been protected with resist.

7. PCs may also be produced by several photographic methods in which the copper-clad surface is coated with a light-sensitive emulsion.

8. The conductor layout should use gradual bends, ample width, proper spacing, and insulated components as cross-overs.

9. Begin the design with components that have common connections.

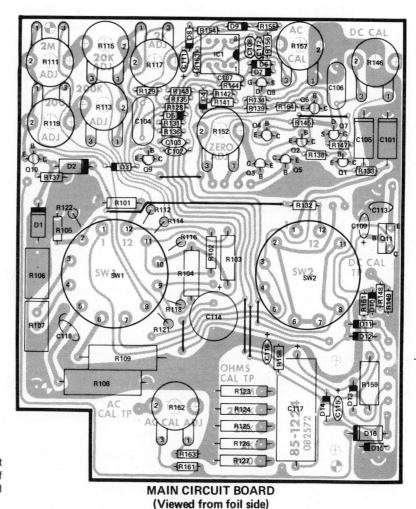

MAIN CIRCUIT BOARD
(Viewed from foil side)

Figure 8-10 Digital meter main circuit board. (Reprinted by permission of Heath Company. Copyright © 1972 all rights reserved.)

Problems **8-1** How much current will a 3.18 mm-wide conductor (2 oz/ft² copper) carry without heating beyond 40°C?

8-2 Determine the required width of a 1 oz/ft² copper conductor needed to carry 3.5 amperes without heating beyond 40°C.

8-3 Calculate the spacing required when two adjacent conductors have a potential difference of 500 volts peak.

8-4 Properly connect the lands in the PC layout of Figure 8-11 to conform to the common-emitter amplifier schematic diagram.

8-5 Prepare a full-scale PC layout of the crystal oscillator circuit shown in Figure 6-1.

8-6 Lay out the PC of the code instructor circuit of Figure 6-5 to full scale. Terminal pads should be provided for the speaker, key, and battery. Assume transformer-mounting dimensions of 2 cm × 5 cm. Mount the potentiometer with its shaft protruding from the copper-clad side of the board.

8-7 Draw the PC of the phonograph amplifier circuit of Figure 6-7 to full scale. Include terminal for connection to the pickup cartridge, speaker, switch, and battery. Mount the potentiometers with their shafts protruding from the copper side of the board.

8-8 Prepare a PC layout of the operational amplifier circuit shown in Figure 6-10 to double scale. The microphone jack, output jack, and batteries are not to be included on the PCB; provide terminal pads for their connections.

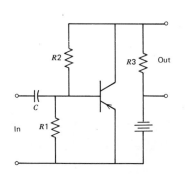

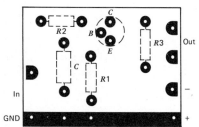

Figure 8-11 Printed circuit problem

Now return to the self-evaluation questions at the beginning of this chapter and see how well you can answer them. If you cannot answer certain questions, place a check next to them, and review the appropriate sections of the chapter to find the answers.

References Charles J. Baer, *Electrical and Electronics Drawing,* Third Edition, McGraw-Hill, New York, 1973, pp. 174–185.

Sal Di Nuzzo, ''Printed Circuit Technology,'' *Electronics World,* October 1969, pp. 37–41.

Phillip C. Hecker, and Charles T. Novak, ''Connectors for PC Boards,'' *Electronics World,* October 1969, pp. 52–55.

Victor Liebmann, ''Conformal Coatings for Printed Circuits,'' *Electronics World,* October 1969, pp. 50–51.

Nicholas M. Raskhodoff, *Electronic Drafting and Design,* Second Edition, Prentice-Hall, Englewood Cliffs, N.J., 1966, pp. 539–576.

Hal R. Roffmann, "High Density PC Boards," *Electronics World,* October 1969, pp. 42–43.

George Shiers, *Electronics Drafting,* Prentice-Hall, Englewood Cliffs, N.J., 1962, pp. 458–468.

Norman A. Skow, "Printed Circuit Laminates," *Electronics World,* October 1969, pp. 44–46.

Donald L. Steinback, "Printed Circuit Kits for Short Runs," *Electronics World,* October 1969, pp. 56–57.

Chapter 9

Integrated Circuit (IC) Drawing

Instructional Objectives
1. To learn the basics of microelectronics.
2. To be aware of the advantages and disadvantages in the use of integrated circuits (ICs).
3. To become aware of the difference between PC and IC drawings.
4. To extend PC drafting methods to IC drawings.
5. To learn about IC masks and how they are to be drawn.
6. To learn how to plan and layout resistive and capacitive IC components.
7. To design a functional IC from a schematic diagram.

Self-Evaluation Questions
Test your prior knowledge of the information in this chapter by answering the following questions. Watch for the answers to these questions as you read the chapter. Your final evaluation of whether you understand the material is measured by your ability to answer these questions. When you have completed the chapter, return to this section and answer the questions again.

1. What is included in the study of microelectronics?
2. Why are IC fabrication methods limited to large production quantities?
3. What is a microelectronic matrix?
4. What specific drafting problems are encountered in preparing layouts for ICs?
5. What is the first step in making IC drawings?
6. Why are microelectronic layouts drawn to a very large scale?
7. How are resistors drawn in an IC layout?
8. Why are expansion cracks in an IC conductor not a problem?
9. How are circuit crossovers accomplished in an IC wafer?
10. In what way has the packaging of IC wafers become standardized?

9-1
Microelectronics

The study of *microelectronics* refers to extremely small electronic components and circuit assemblies, made by *thin-film, thick-film,* or *monolithic* techniques. The advantages of microelectronics surpass those given in Section 8-1 for PCs, but their application is still further limited to large production quantities. Thin- and thick-film circuits may be considered as extensions of the earlier discrete-component electronics in that they are assemblies of components either formed on or fixed onto the surface of an insulating *substrate*.

In thin-film circuits, passive components such as resistors, capacitors, and interconnection wiring are formed directly on the surface of glass or ceramic *substrates* using *evaporation techniques*. Semiconductor components, such as diodes and transistors, are fabricated as separate semiconductor wafers and assembled into the circuit.

Thick-film circuits use the silk-screen process (Section 8-3) to form conducting lead patterns and passive components. *Alumina,* about 1.27 cm (0.5 in.) square and 150 mm (0.06 in.) thick, is a typical substrate. A *metallized ink* interconnection pattern is silk-screened on the substrate and then fired at a temperature of about 700 °C. A metal-glass *slurry* is silk-screened and similarly fired to form resistors. Capacitors are either miniature discrete components soldered into the circuit or film-type fabricated on the substrate. The dielectric in the film type may consist of a mixture of glass and ceramic applied as a paste and fired. Active devices are in the form of single-sided *planar* transistor and diode wafers or pellets fused directly to the lead patterns.

The silicon *monolithic* IC is now generally the preferred approach to microelectronics. The active and passive elements are all formed on the substrate of an *intrinsic* or pure silicon slice by *diffusing* impurities into selected regions to change their electrical characteristics; this is basically a *diffused planar process* (Section 9-2). Contact regions and interconnections are formed or metallized upon the active and passive areas to form the complete electronic circuit (Fig. 9-1).

The same circuit is repeated as many as 500 times on a single 4-cm-diameter silicon slice; all of the circuits are processed at the same time. Each IC wafer is about 1.3 mm square and may contain up to 50 circuit elements.

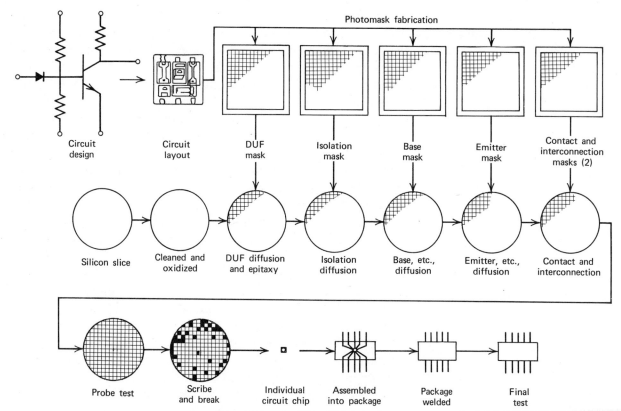

Figure 9-1 Monolithic IC fabrication sequence. (Courtesy of Texas Instruments Incorporated.)

**9-2
The Planar
Process**

The planar process involves a combination of oxidation, selective oxide removal, and impurity diffusion of a silicon slice. *Silica* or *silicon dioxide* is reduced by heating with coke in an electric furnace. The resulting silicon must be repeatedly crystallized to obtain the required high purity. A carefully controlled impurity is added to the pure silicon to obtain *n*-type or *p*-type silicon. The single crystal *grown* ingots are sawed into slices, lapped, and polished to a thickness of about 0.25 mm.

The impurity may be selectively added to certain areas of a pure silicon slice by heating the slice in the impurity *dopant* vapor; a high concentration of dopant is deposited on the silicon surface. The slice is then heated at a carefully controlled higher temperature so that the dopant atoms move or diffuse into the silicon. The depth of diffusion is controlled by both temperature and time.

The diffusion process may be controlled by forming a silicon *mask* on the silicon slice to prevent diffusion into selected areas. The silicon slice is oxidized by heating it in a flow of oxygen. The oxide is then removed from selected regions by etching with *hydrofluoric acid* (or another appropriate etchant) as shown in Figure 9-1. Resist is applied photographically similar to the method described in Section 8-2. The steps involved in selective oxide removal are shown in Figure 9-2.

The basic processes of *silicon dioxide* masking, etching, and diffusion with *n*-type or *p*-type dopant are progressively accomplished (Fig. 9-3) to produce a MOS/FET transistor. Note that pure silicon dioxide is used to selectively insulate the source,

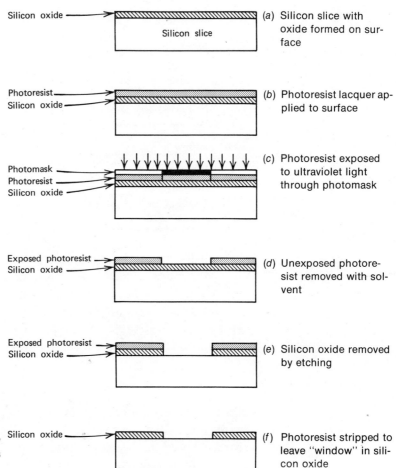

Figure 9-2 Selective oxide removal by photoresist process. (Courtesy of Texas Instruments Incorporated.)

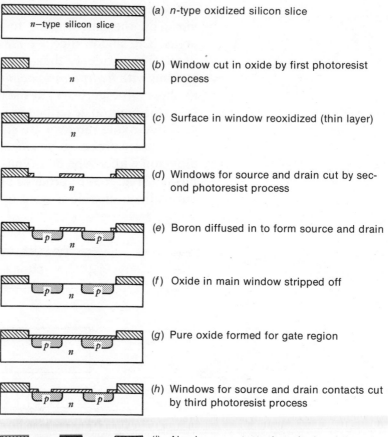

(a) *n*-type oxidized silicon slice

n—type silicon slice

(b) Window cut in oxide by first photoresist process

n

(c) Surface in window reoxidized (thin layer)

n

(d) Windows for source and drain cut by second photoresist process

n

(e) Boron diffused in to form source and drain

p   *p*
n

(f) Oxide in main window stripped off

p   *p*
n

(g) Pure oxide formed for gate region

p   *p*
n

(h) Windows for source and drain contacts cut by third photoresist process

p   *p*
n

Figure 9-3 Steps in MOS FET fabrication. (Courtesy Texas Instruments Incorporated.)

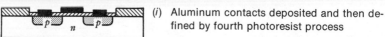

(i) Aluminum contacts deposited and then defined by fourth photoresist process

p   *p*
n

drain, and channel regions except where the aluminum contacts are to be made. Aluminum is vaporized in an evacuated chamber; the vaporized aluminum then deposits on areas not protected by a mask or resist. Since diodes are also fabricated from *p*-type and *n*-type material, they are prepared at the same time the transistors are formed on the silicon slice.

**9-3
Integrated
Circuit
Resistors** Intrinsic silicon is a good insulator and therefore has a very high resistance. Since the resistivity of silicon depends on the concentration of its diffused impurity, most resistors are formed at the same time as the *p*-type transistor base region. The concentration of the impurity is determined by the require-

ments of the transistor; therefore, the dimensions of the resistor must be designed to give the required resistance value. Resistance is directly proportional to conductor length and inversely proportional to the area of the conductor. Typical resistance of a stripe of p-type base material 0.025 mm (1 mil) wide and 0.025 mm long is 100 ohms. Therefore a stripe 0.25 mm (10 mil) long and 0.025 mm wide will have a resistance of 1000 ohms. High values of resistance are obtainable if the long stripe is arranged back and forth in a grid form. If the stripe is made wide and short, resistances may be fabricated to values as low as 20 ohms. If resistance values down to 2 ohms are needed, the higher concentration emitter diffusion can be used to form n-type resistors.

9-4 Integrated Circuit Capacitors Since a capacitor consists of two metallic plates separated by an insulator or dielectric, capacitors are easily added to ICs. Capacitance is directly proportional to the area of the plates and inversely proportional to the thickness of the dielectric. Since space is limited on an IC wafer, capacitance is usually limited to a few hundred picofarads.

The bottom plate could be an n-region diffused into the silicon at the same time as the transistor emitter is diffused. A controlled thickness of silicon dioxide or *titanium dioxide* is formed on the surface of the n-region to serve as the dielectric. The top plate may consist of aluminum deposited at the same time as the interconnection pattern.

Capacitance may be approximated with the formula $C = 8.85 \times 10^{-3} KA/d$ pF in which K is the dielectric constant, A is the area of the plate in square millimeters, and d is the thickness in millimeters.

A reverse-bias pn transistor junction contains few charges between the p and n regions, and therefore acts as a capacitor. A small capacitor can be formed at the same time as the other elements with no additional processes. Since the capacitance of a pn junction depends on the value of reverse voltage, it is necessary to arrange the correct voltage bias in the circuit. Furthermore, to obtain the correct bias, a voltage-dropping resistor stripe may be needed.

The complete IC of Figures 9-4d and 4e contains an example of four types of components that may be included. Coils or in-

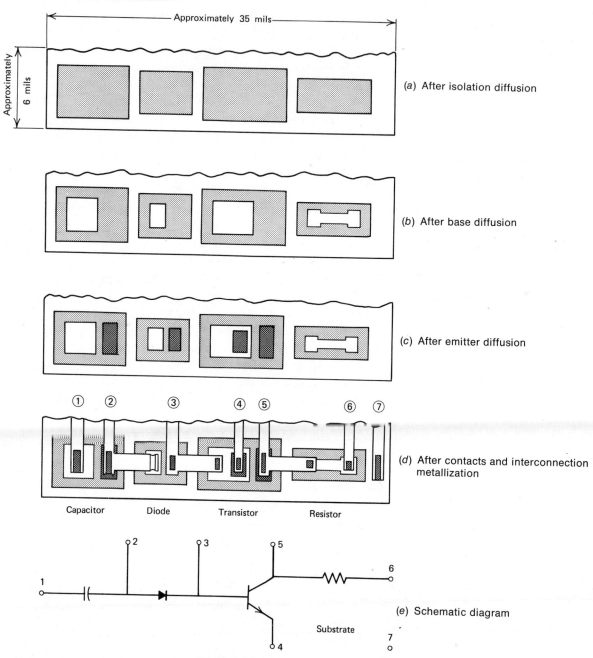

(a) After isolation diffusion

(b) After base diffusion

(c) After emitter diffusion

(d) After contacts and interconnection metallization

Capacitor   Diode   Transistor   Resistor

(e) Schematic diagram

Substrate

Figure 9-4 Integrated-circuit process steps. (Courtesy Texas Instruments Incorporated.)

ductors cannot be included on an IC wafer because of the space limitations; coils must be externally connected to the integrated circuit module. Compare the schematic of Figure 9-4e with the IC layout of Figure 9-4d.

9-5 Integrated Circuit Layout Experience in PC design is of considerable value in IC layout, and crossovers no longer present a problem. It is only necessary to deposit an insulating film of silicon dioxide before the crossover interconnection is made. Drawings are made to a scale of about 150 times actual size; for example, if the wafer is to be a typical 1.3 mm square, then the detailed drawings should be about 20 cm square. As many as eight additional mask drawings may be needed; alignment of the masks with the master drawing is critical. The following steps are required in the order given.

1. Determine the standard packaging (Section 9-6) to be used; this dictates the overall size of the wafer. Refer to Figure 9-5c for the standard 14 pin package dimensions.
2. Begin the master layout about 150 times full scale using grid paper for convenience. Allow additional space on each side for registration marks.
3. The standard packaging determines the location of the terminal lands of the wafer as shown in Figure 9-5a.
4. Study the schematic diagram to find the components that are directly connected to the common ground land.
5. Similarly determine the components that are directly connected to the common positive supply voltage land.
6. Determine the compatibility of components from a study of the schematic diagram.
7. Sketch in components with point-to-point connections using only horizontal and vertical lines.
8. Calculate the areas needed for resistors and capacitors. Record width, length, effective areas, resistance, and capacitance in a table.
9. When pn junction capacitors are specified, arrange them near a correct voltage bias source.
10. Locate the remaining components and lay out their interconnections. Since the lengths of interconnections are very short, expansion cracks are not a problem; interconnections may therefore have 90° bends.

(a) Internal connections

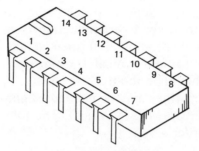

(b) Top pictorial view

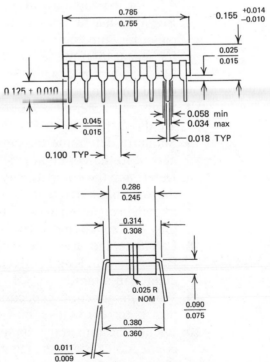

(c) Orthographic drawing

Figure 9-5 Typical dual in-line integrated circuit package.

11. Using the master drawing as an underlay for perfect alignment, produce mask drawings for the processes of isolation diffusion; base, resistor, and diode diffusion; emitter diffusion; pre-ohmic etch for contacts; and metalization for interconnections.

12. Sufficient duplicates of all masks are made photographically, arranged in a matrix, and reduced to full scale. These masks are used to process as many as 500 chips at the same time from a single silicon slice.

9-6 Packaging The TO-5 cylindrical container is used for single transistor packaging and also for the similar TO-100 the containment of the first IC designs (Appendix E). Only ten leads can emerge from the bottom of this container. Since their leads are fragile, socket insertion or soldering into a printed circuit is difficult. Therefore, the *dual in-line package* (DIP) was developed and standardized to overcome the TO-100 limitations. The 14-pin DIP shown in Figure 9-5 is the most popular; it is available in either a plastic or ceramic housing. The pin spacing of 0.100 in. (2.54 mm) is also used in the 12, 16, 24, 36, and 42 pin DIPs. The name dual in-line package infers that two circuits are contained in one package; a single circuit 8-pin *mini-DIP* is also available.

In most cases, manufacturers adhere to a standardized pin-numbering system; the manufacturers' data sheets should be consulted for possible variations. Most DIPs locate pin 1 at the upper left-hand corner of the top view. The upper end is indicated by a *notch* or a *dot* at pin 1 as illustrated in Figure 9-5*b*. Sockets and *molex strip* contacts are readily available in the standard pin spacing. Resist appliqués are also obtainable with the standard land spacing for preparation of PC boards (see Fig. 8-2).

9-7 Drawing ICs in Linear and Digital Circuits Diode-transistor logic (DTL) digital circuits have been standardized. Figure 9-6 shows a typical DTL simple-gate IC known as a dual 4-input NAND gate (SN15830N). See Section 3-5 and Figure 3-8*e* for review of a single 2-input NAND gate. The NAND gate has a 1-level output under all input conditions *except* when *all* inputs are at a 1 level.

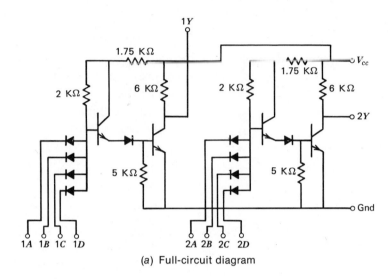

(a) Full-circuit diagram

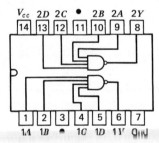

(b) Logic and package diagram

Figure 9-6 Typical DTL NAND gate IC. (Courtesy of Texas Instruments Incorporated.)

Compare the location of the two sets of four diode inputs in the full-circuit diagram Figure 9-6a with the inputs in the Figure 9-6b package diagram. Furthermore, the 1Y and 2Y outputs of the NAND gates are found to be pin 6 and 8, respectively, in the package diagram.

The operational amplifier shown in Figure 9-7 is a linear circuit in which the output is proportional to the input signals at all times. An operational amplifier consists of two *differential amplifiers* followed by a *driver stage* and a common-collector output stage. Since a differential amplifier serves to amplify the difference between two input signals, the OP AMP will contain two inputs and one output as shown in Figure 9-7. Review Section 5-9 and Figure 5-8a.

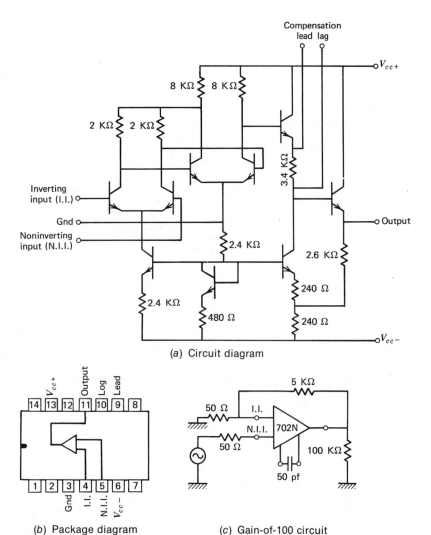

(a) Circuit diagram

(b) Package diagram (c) Gain-of-100 circuit

Figure 9-7 General purpose operational amplifier (SN72702N) (Courtesy of Texas Instruments Incorporated.)

Pins 9 and 10 of the Figure 9-7b package diagram are provided to allow external compensation components to be added. The 50-pF capacitor shown in Figure 9-7c ensures stable operation under various feedback conditions.

The Figure 9-8 basic calculator circuit contains an 18-pin MM5736 calculator IC and a 14-pin 75492 digit driver IC. Notice the similarity of this circuit to the digital-readout display circuit of Figure 5-8d. The keyboard pushbutton switches provide three inputs (K1, K2, and K3) to the NM5736 calculator

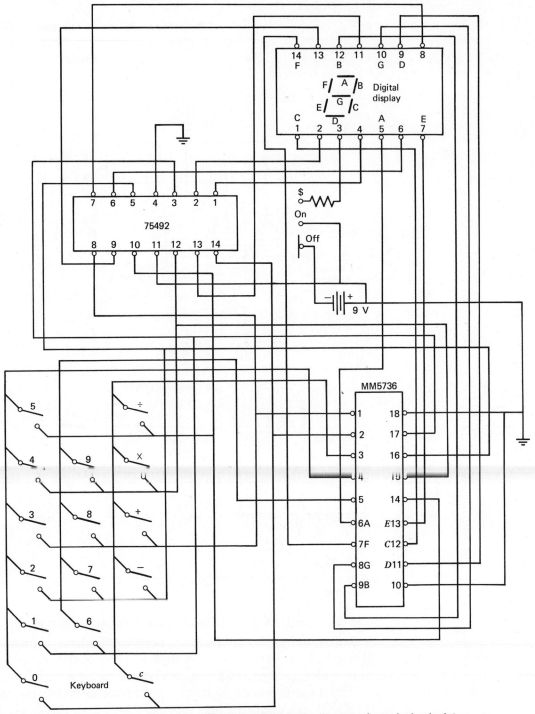

Figure 9-8 Schematic diagram of a typical calculator.

IC. One set of calculator IC outputs from pins 1, 2, 14, 15, 16, and 17 drive the 6-digit LED display. The other set of outputs (A through G) provides a 7-segment coded output for the calculator's display register.

Summary
1. Microelectronics includes the assembly of extremely small, active and passive electronic components into ICs.
2. IC modules have advantages of miniaturization, reduction of interconnection expense, and reliability.
3. The use of IC modules is limited to large production runs because of the high initial production costs.
4. Usually 500 IC wafers are processed at the same time from a single silicon slice.
5. The popular monolithic planar design approach involves the deposition and diffusion of substances on a silicon substrate to form electronic components and their interconnections.
6. In the planar process, a separate mask is prepared for the selective deposition of each n-type or p-type impurity.
7. Accurate masks are necessary for the deposition of insulating silicon dioxide and interconnecting metals.
8. The layout of an IC is similar but to a much different scale than a PC.
9. The industry has standardized on the dual in-line package (DIP) with 2.54 mm pin spacing.

Problems
9-1 What length of resistive element is necessary to produce a 470 ohm resistor using p-type base material to a 0.075 mm width?

9-2 A 2200 ohm IC resistor is to be deposited on a silicon substrate from p-type base material. What width must be specified if the length cannot exceed 0.40 mm?

9-3 Calculate the capacitance when two 0.20 mm × 0.20 mm aluminum plates are separated by a 0.010 mm-thick dielectric. Assume a dielectric constant $K = 8$.

9-4 Determine the thickness of a titanium oxide dielectric with constant $K = 176$, which is needed to obtain a 306 pF capacitance. Assume the plates are each 0.10 mm × 0.10 mm.

9-5 Design and draw the IC for the schematic diagram shown in Figure 9-9a. Assume that resistor material 0.025 mm ×

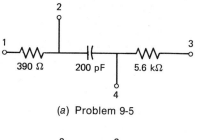

(a) Problem 9-5

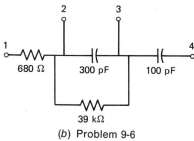

(b) Problem 9-6

Figure 9-9 Schematic diagrams for IC Problems.

0.025 mm has a resistance of 100 ohms. Use a dielectric constant $K = 8$ in the design of the capacitor.

9-6 Design and draw the IC for the schematic diagram (Fig. 9-9b). Use a dielectric constant $K = 176$ for titanium oxide, and 100 ohm resistance for a 0.025 mm × 0.025 mm square.

9-7 Convert the schematic diagram shown in Figure 8-11 to an IC layout. Two terminal pads each should be provided for the input, output, and battery. All components must be contained on a 1.3 mm × 1.3 mm wafer. Draw circuit to 150 times full scale.

9-8 Draw all of the necessary masks needed to produce the IC circuit designed in Problem 9-7.

Now return to the self-evaluation questions at the beginning of this chapter and see how well you can answer them. If you cannot answer certain questions, place a check next to them, and review the appropriate sections of the chapter to find the answers.

References **Charles J. Baer,** *Electrical and Electronics Drawing,* Third Edition, McGraw-Hill, New York, 1973, pp. 185–206.

Robert Boylestad, and **Louis Nashelsky,** *Electronic Devices and Circuit Theory,* Prentice-Hall, Englewood Cliffs, N.J., 1972, pp. 497–519.

R. G. Hibberd, *Integrated Circuits,* McGraw Hill, New York, 1969.

Chapter 10

Pictorial Assembly Drawings

1. To learn the advantages of pictorial assembly drawings.
2. To combine and extend previous discussions of components and their wiring into complete functional electronic assemblies.
3. To understand that the design of an electronic device depends on many factors beyond its circuitry.
4. To learn the difference between isometric, oblique, dimetric, and perspective pictorial drawings.
5. To understand the method and purpose of foreshortening in oblique and dimetric drawing projections.
6. To draw isometric circles and/or ellipses proficiently.
7. To make isometric pictorial assembly drawings proficiently.
8. To learn to draw one- and two-point perspective drawings of electronic components.

Self-Evaluation Questions Test your prior knowledge of the information in this chapter by answering the following questions. Watch for the answers as you read the chapter. Your final evaluation of whether you understand the material is measured by your ability to answer these questions. When you have completed the chapter, return to this section and answer the questions again.

1. How are pictorial drawings used in the assembly of an electronic device?
2. In what way is the sales department of an electronic manufacturer concerned with an electronic design?
3. What is an isometric drawing?
4. What advantage does an isometric drawing have over an oblique drawing?
5. What is an oblique drawing?
6. What is a dimetric drawing?
7. Why is a dimetric drawing more difficult to draw than an isometric drawing?
8. Under what circumstances is an oblique drawing preferred to an isometric drawing?

9. What is a perspective drawing?
10. What disadvantage does a perspective drawing have when compared with other types of pictorial drawings?
11. How many and which triangles are required to make an isometric drawing?
12. How many true edge dimensions are there in the dimetric drawing of a cube?
13. How is a cylindrically shaped component such as a potentiometer drawn in isometric view?
14. What is a *call-out* and how is it used in electronic drafting?

**10-1
Purpose of
Pictorial
Drawings**

When an electronic component or an assembly of electronic components is produced by a manufacturer, it should be obvious that several kinds of drawings are necessary to convey the particular information demanded by the many production departments involved in its manufacture. Precise instructions must be given to the production workers who build or assemble the components to the chassis. These workers may have little or no training in electronics or orthographic print reading. Therefore, construction and assembly drawings must be clear. Since orthographic drawings (Section 1-11) are difficult for untrained workers to understand, pictorial drawings are usually supplied. Pictorial assembly drawings are also found useful by the production control, time and motion studies, quality control, inspection, service, and sales departments. Many electronics students are aware of the advantages of the pictorial assembly drawings supplied by kit manufacturers. The three-dimensional effect simulated in pictorial drawings is achieved through such drawing methods as *isometric, oblique, dimetric,* and *perspective*.

**10-2
Isometric
Drawing**

An *isometric* view of a cube is approximated when it is tilted toward the viewer so that three surfaces and nine edges are visible (Fig. 10-1). Isometric drawings may also be shown with the object tilted upward so as to illustrate an important bottom feature; likewise, the view may be tilted to the right or the left. In nearly all applications of the isometric system, the lengths of *all* edges are drawn to the *same* scale. While this system conveys a realistic but *not* a photographic impression of the object, it has the advantage that line dimensions may be scaled

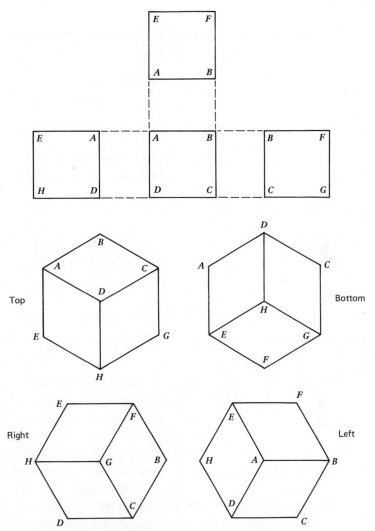

Figure 10-1 Cube in common isometric views.

directly from an isometric drawing. Dimension, hidden, section, and centerlines are *not* shown as a rule.

The following isometric drafting procedure is recommended and illustrated in Figure 10-2.

1. Outline an irregular object such as the AF choke as a box whose dimensions are the same as the choke's major dimensions.
2. Decide on the tilt direction and the leading edge on the basis of relative importance of a view or edge.
3. Draw the leading edge of the "box" vertical and to scale.

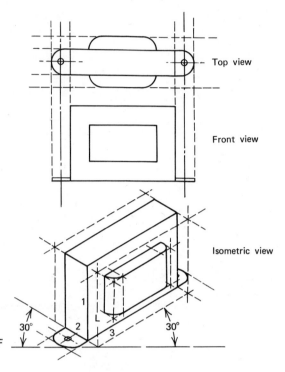

Top view

Front view

Isometric view

Figure 10-2 Isometric drawing of an AF choke.

4. Draw *projectors* at 30° angles to the horizontal from the bottom of the leading edge line (L).

5. Lay out the length and width of the "box" on the projectors.

6. Lay out the core width along one projector from the core corner to scale. This will determine the position of line 2 in Figure 10-2 and establish a new origin.

7. Measure and lay out the core height (1) vertically to establish the actual leading edge.

8. Draw the actual core length (3) along the 30° projectors.

9. Using 30° and vertical lines, complete the core projection; the length of all edges should be to scale.

10. The location and dimensions of the coil is similarly produced from its boxlike layout to scale.

11. The semicircular coil ends are drawn as *quarters of an ellipse* using a 30° ellipse template.

12. Similarly the choke's mounting lugs are drawn as extensions from the bottom of the core.

The most difficult part of any pictorial drawing is the conversion of circles into appropriate shapes with circular dimensions. When objects having cylindrical or conical shapes are placed in an isometric position, the circles are viewed at an angle and appear as ellipses. If an appropriate ellipse template is not available, the construction method shown in Figure 10-3 produces satisfactory results.

Construction of an appropriate isometric ellipse (Fig. 10-3):

1. Locate O on the centerline origin of the circle.
2. Draw $A'B'$ and $C'D'$ at 30° to the centerline, and through the c origin O.
3. Use original circle radius R to locate points A, B, C and D; $R = OA = OB = OC = OD$.
4. Locate X by drawing a line through C at 60° to the horizontal centerline; point E is found at the intersection with the horizontal centerline.
5. Similarly locate center points Z and F.
6. Use points E and F as centers to draw *minor arcs* $\overline{AC}$ and $\overline{DB}$.
7. Use points X and Z as centers to draw the *major arcs* $\overline{BC}$ and $\overline{AD}$.

In order to draw right circular cylinders such as potentiometers, sketch both ellipses even though part of the view is hidden by intervening material. The cylinder height will fall along 30° projectors as shown in Figure 10-4. You may find it worth-

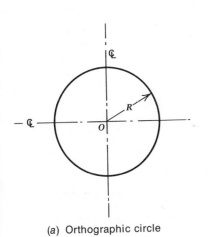

(a) Orthographic circle

(b) Location of centers for major and minor arcs

(c) Drawing major and minor arcs

Figure 10-3 Construction of an isometric ellipse.

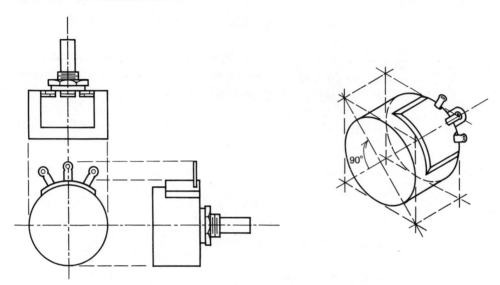

Figure 10-4 Orthographic and isometric drawing of a potentiometer.

while to review the pictorials of components in Chapter 4. An important rule about the right circular cylinder is that the longest or *major diameter* of the ellipse always appears perpendicular to the cylinder axis.

10-3 An *oblique pictorial* drawing is the least difficult to draw com-
Oblique pared to all other pictorial methods. The front view appears in
Drawing true dimensions, parallel to the *picture plane,* and thus may be directly drawn as shown in Figure 10-5. Another advantage of oblique drawing is that if the side containing circles, curves, or contours is chosen as the front view, then elliptical construction is unnecessary. Circles on the remaining top or side views, however, must be converted to ellipses (see Fig. 10-5).

Construction of an oblique drawing (Fig. 10-5):

1. Choose the front view on the basis of importance and the view containing the most curves and/or holes. It is also desirable to choose the view having the longest dimension so as to reduce distortion.
2. Directly transfer the front view in true scaled dimensions.
3. Determine whether the top view or side view is more important.

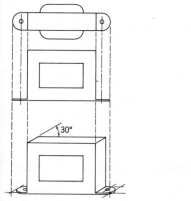

Figure 10-5 Oblique drawing of an AF choke.

4. The box construction described in Section 10-2 may be used if the object has an irregular shape.
5. If the top view is important, draw 60° oblique projectors for the receding views from all corners; draw 30° projectors when the side view is important.
6. Cut off receding projector lines so that the depth appears correct. The lines may be full length, but a more natural appearance results if they are *foreshortened* or reduced to three-quarters or one-half size.
7. Connect the ends of the foreshortened projectors with horizontal or vertical lines to complete the projection.
8. Locate centerlines for all holes or circular shapes.
9. Convert any circles contained in the top or side view to ellipses with a template or the method of Section 10-2.

10-4
Dimetric
Projection
Pictorial drawings using *dimetric projection* are less distorted than those drawn by isometric or oblique projection but are more difficult to draw. The construction steps are similar to those for an isometric drawing as outlined in Section 10-2 except that the projectors are drawn at 7° and 41° (tangents of ⅛ and ⅞, respectively) instead of 30° each. Usually the front-view dimensions are in the same ratio (1:1) while the third or receding axis (3) is foreshortened by using a half-scale (½:1). Since two different scales and angles are used when laying out a dimetric drawing, circular parts are much more difficult to draw. Figure 10-6 shows the choke drawn in dimetric projection.

10-5
Perspective
Drawing
Only an *approximate* representation of an object (as it appears to the eye) is obtained using the above-described methods of pictorial drawing. A *perspective drawing* represents an object as it appears to an observer or camera stationed at a particular position relative to it. If, for example, an observer is standing in the center of a pair of railroad tracks, the tracks appear to converge to a point at the horizon. The tracks may be considered to be projectors and the point of convergence is known as the *vanishing point* in a *one-point perspective* drawing.

Construction of a *one-point* perspective drawing of a cube:

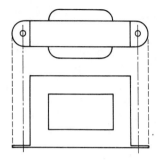

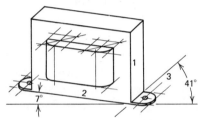

Figure 10-6 Dimetric projection of an AF choke.

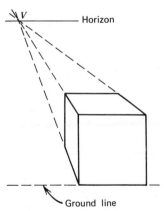

Figure 10-7 One-point perspective of a cube.

1. Draw the front view of the cube as would be done in oblique or orthographic drawing (Fig. 10-7).
2. Choose a vanishing point *V* to the right or left of the front view and near the top of the sheet.
3. Draw light receding lines from the corners of the front view to the vanishing point *V*.
4. Locate the receding corners of the cube by a foreshortened length (½:1 in Fig. 10-7) along the receding lines for best appearance.
5. Complete the cube with a horizontal and a vertical line connecting the receding corners.

The dimensions and shape of the front view are true; circular objects on the front face are drawn as true circles unless they project from the face a significant distance as shown in Fig. 10-8. Direct measurements cannot be made of the receding or rear edges. In the *two-point perspective* shown in Fig. 10-9, only the front edge lines marked 1, 2, and 3 are true dimensions. It will be observed that the foreshortening in a two-point perspective is variable and dependent upon the positions of the vanishing points.

Construction of a *two-point* perspective drawing (Fig. 10-9):

1. Draw the vertical edge (1) to scale from a horizontal ground line.
2. Choose two vanishing points on a horizontal line near the top of the sheet. If both vanishing points are at an angle of 45° to the ground line from the bottom of the vertical edge, the two vertical faces will receive equal treatment. When it is desirable to emphasize one of the faces, one point may be located at an angle of 30° to the ground line while the other point is at a 60° angle.
3. Draw projectors from the top and bottom of the vertical edge (1) to both vanishing points.
4. Draw the core depth (2) to scale along the projector to the left vanishing point.
5. Draw the core length (3) to scale along the projector to the right vanishing point.
6. Erect vertical lines 4 and 6 from the receding ends of lines 3 and 2 to the top projectors.
7. Connect the top of line 6 to the right vanishing point; this will establish the length and direction of edges 7, 8, and 9.

8. Similarly establish the coil location using the box method of Section 10-2.

9. The quarter-circles of the coil are projected on the perspective plane as ellipses similar to the method in Figure 10-3. A perspective square is drawn and diagonals are put in to locate the center; the sides are bisected. The ellipse will be correct if it is tangent to the centers of the sides of the enclosing perspective rhombus.

**10-6
Photographic
Pictorials** Pictorial drawings are difficult to produce; therefore, many electronics manufacturers resort to *photographic pictorials* (Fig. 10-10). This procedure is labor saving since most of the components of an assembly are illustrated. However, it is extremely difficult to trace the wiring of an assembly from such photographs unless much photo retouching is done. The photo density of the wires may be changed on the negative so that the wires are more visible on the positive print. Since a photograph contains few white areas unless retouched, component designations are not lettered on or adjacent to the component. *Call-outs,* or circled designation codes connected by arrows to the component, are usually arranged in logical order with few crossing arrows on the periphery of the photograph. Three such call-outs are shown in Figure 10-10. The radio and television service industry widely uses photo pictorials for component location.

**10-7
Pictorial
Assembly
Design
Factors** The electronic draftsman is usually introduced to a proposed electronic assembly by way of a *bread-boarded* circuit approved by the experimental design engineer. Components are fastened to a flat board, laminated plastic sheet, or a salvaged chassis but not necessarily in the best space-saving position. Components have been mounted for convenience in wiring and possible replacement. The initial problem is to convert the bread-boarded circuit to a schematic diagram, as discussed in Section 6-3.

A duplicate set of loose parts are usually obtained so that the best space-saving arrangement may be determined. The actual components used in the bread-boarded circuit must be checked against the manufacturer's specification sheets for overall and

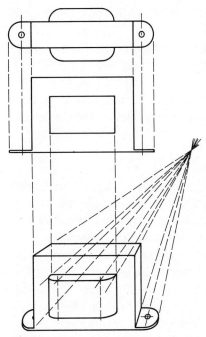

Figure 10-8 One-point perspective drawing of an AF choke.

mounting hole dimensions and tolerances. Sometimes the specifications supplied by the design department may be so rigid that the identical component must be used, although it is unwise not to be aware of acceptable substitutions.

An *assembly drawing* must illustrate the physical arrangement and mechanical details of components and hardware grouped together as a unit. Such a drawing, therefore, should give an accurate description of the shape of each component and how and where they are mounted (Fig. 10-11). Usually several components are combined to form a unit assembly or *subassembly*. A number of such subassemblies using suitable fastener hardware are then fastened together to form the com-

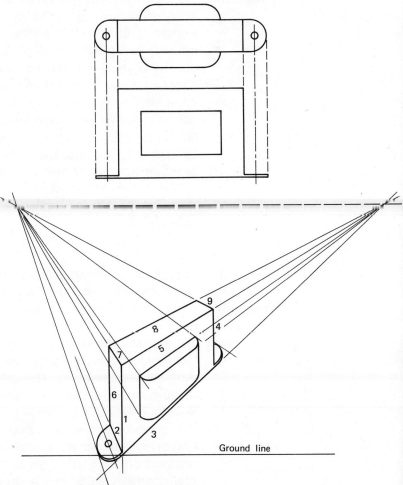

Figure 10-9 Two-point perspective drawing of an AF choke.

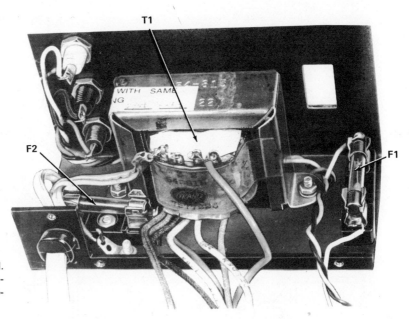

Figure 10-10 Chassis photo-pictorial. (Reprinted by permission of Heath Company. Copyright © 1972 all rights reserved.)

plete assembly. It is important that nothing in the assembly is left to the interpretation of the assembler. Be complete and explicit, although needless repetition of small details, such as fasteners, can be avoided by giving an inset view and local notes.

Before proceeding with actual component placement, it is advisable to check with the engineer responsible for the circuit design, particularly with respect to magnetic, electrostatic, or capacitive coupling; voltage and current ratings; and shielding requirements. Since conservation of space is not considered when bread-boarding a circuit, the presence of high voltages or some of the above factors may dictate a larger or insulated chassis.

The sales department is especially interested in the physical appearance, size, and shape of the completed assembly and should therefore be consulted. There have been many instances where a new product is developed as a result of the sales department's recommendation for a competitive item or a new product to complete the line. The particular chassis assembly may be a part of a larger unit or cabinet. For example, automotive test equipment may consist of many portable units of the same physical size so that the automobile mechanic may

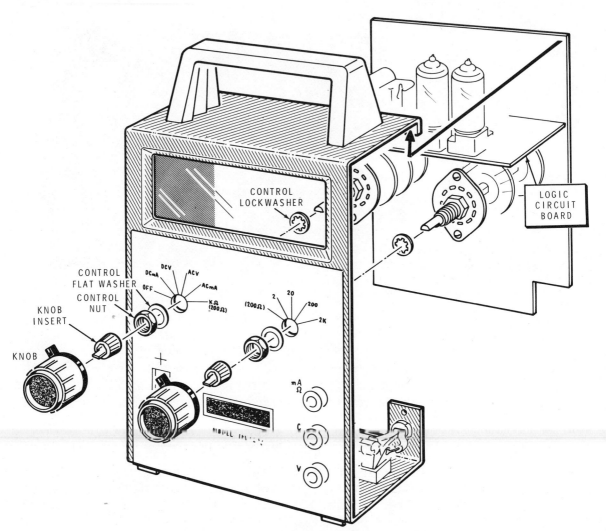

Figure 10-11 Hardware assembly of a digital multimeter. (Reprinted by permission of Heath Company. Copyright © 1972 all rights reserved.)

easily store the tester in its cabinet; the contents of the individual test units will then have no relationship to the chassis volume. The dimensions and packaging of the chassis may also be determined by styling. Transistor radios are often designed to fit a coat pocket, although the same chassis components may be enclosed in a much larger container, as in a table radio.

The service department may submit different recommendations from the sales department regarding chassis size, since accessibility of parts and test points is their main interest. It is important that components having a short life expectancy (such as batteries) are mounted for easy replacement; plug-in components are often specified for this purpose if cost is not a major consideration. The electronic service industry is now advocating the use of IC sockets instead of the permanent solder-mounting of ICs to facilitate IC replacement.

Some electronic components may be damaged or become unreliable when subjected to vibration; to reduce such service problems, the chassis mounting of such components is given special consideration. As an example, an automobile radio is subject to more vibration than a domestic table radio. Choice of components and their mounting is also dependent upon operating conditions, particularly humidity and temperature. As an extreme case, imagine the design problem of building a television camera to be landed and operated by our astronauts on both the cold and hot sides of the moon.

After all design factors are given full consideration, tentative orthographic drawings are made from preliminary sketches. All mounting holes are completely dimensioned and tolerances indicated. It is advisable that a *prototype* chassis is fabricated and assembled to the specifications of these initial preliminary drawings. Assuming that all components fit the chassis, the prototype assembly is then submitted to the circuit designer and test section for extensive testing. The results of these tests are to confirm the original design calculations, to determine where changes are needed, and to provide a firm basis for further redesign and development. For simple designs, the prototype may serve as a production model.

The service department may wish to include a simplified version of the orthographic chassis layout in the service manual to help the purchaser locate components. The drawing must be to scale, but dimensions and wiring are seldom necessary.

Details of the mechanical and electrical features may now be taken from the prototype and the orthographic chassis layout to prepare the pictorial assembly drawings. Choose the viewing or display angle of the pictorial drawing to show as many components as possible. The terminals of the components should be visible for wiring purposes as shown in Figure 10-12; when this is not feasible, a partial inset view may be included, or dashed

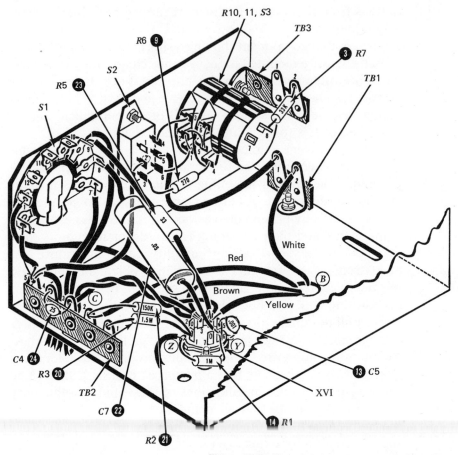

Figure 10-12 Pictorial view of chassis wiring. (Courtesy of EICO Electronic Instruments Co., Inc.)

lines may be drawn for hidden details. Dimensions and hidden lines are rarely required; a hidden part is sometimes moved slightly out of its true scale position to prevent one component from obscuring another important part or terminal.

Summary 1. Pictorial assembly and wiring drawings are needed when assemblers are not trained in orthographic print reading.
2. Oblique and isometric pictorial drawings do not quite show the true shape of an object or assembly but are the easiest to draw.
3. An isometric view of an object is obtained when it is tilted symmetrically to show three surfaces.

4. Oblique projection is the only method that includes one view to scale and true shape.

5. Two-point perspectives are considerably more difficult to draw than oblique and isometric projections but are (almost) photographically realistic.

6. Photographic pictorials contain too much detail for rapid location of components and wiring and therefore may require considerable touch-up and many call-outs.

7. Proper electronic assembly design must consider the compatibility of electronic components, simplicity of mechanical assembly, accessibility for servicing, and the additional requirements of the sales and service departments.

Problems **10-1** Make an oblique, isometric, dimetric, and one- and two-point perspective drawing of a 4 cm cube; include all construction lines.

10-2 Prepare an oblique, isometric, dimetric, and one- and two-point perspective drawing of a 4 cm diameter and 5 cm high right circular cylinder.

10-3 Obtain an SPDT slide switch from your instructor and draw an isometric view to triple scale (Fig. 4-5b).

10-4 Make a one-point perspective drawing of an SPST toggle switch (Fig. 4-5a) to triple scale.

10-5 Produce a two-point perspective drawing of a power SCR (Fig. 4-8) four times actual size.

10-6 Make an oblique pictorial drawing of a small speaker to double scale (Fig. 4-11c).

10-7 Determine which two objects in Figure 10-13a are the same.

10-8 Which of the objects in Figure 10-13b are the same?

10-9 Find the identical objects in Figure 10-13c.

10-10 Design the assembly of a preamplifier (Fig. 6-4) mounted in a small plastic case. Terminal strips should be used to mount the small components. The jacks and switch are to be mounted on the plastic case.

10-11 Produce an assembly drawing of the amplifier shown in Figure 6-7, including the PCB developed in Problem 8-7.

10-12 Design and draw the assembly of the PCB prepared in Problem 8-8 of the operational amplifier (Fig. 6-10) in a plastic case.

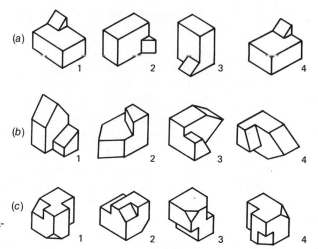

(a)

(b)

(c)

Figure 10-13 Pictorial view identification problem.

Now return to the self-evaluation questions at the beginning of this chapter and see how well you can answer them. If you cannot answer certain questions, place a check next to them, and review the appropriate sections of the chapter to find the answers.

References **Charles J. Baer,** *Electrical and Electronics Drawing,* Third Edition, McGraw-Hill, New York, 1973, pp. 24–31.

Thomas E. French and Charles J. Vierck, *Engineering Drawing,* Tenth Edition, McGraw-Hill, New York, 1963, pp. 157–183.

George Shiers, *Electronic Drafting,* Prentice-Hall, Englewood Cliffs, N.J., 1962, pp. 81–122.

Chapter 11 Electrical Building Construction Wiring Diagrams

Instructional Objectives

1. To understand the difference between schematic and single-line diagrams.
2. To learn to read single-line diagrams.
3. To develop familiarity with electrical symbols used in architectural plans.
4. To learn how basic lighting circuits are wired.
5. To become aware of some basic requirements of the National Electrical Code (NEC).
6. To learn to determine the wire size needed under different load conditions.
7. To develop some proficiency in the design of and drawing of residential wiring circuits.

Self-Evaluation Questions

Test your prior knowledge of the information in this chapter by answering the following questions. Watch for the answers as you read the chapter. Your final evaluation of whether you understand the material is measured by your ability to answer these questions. When you have completed the chapter, return to this section and answer the questions again.

1. How does an architectural wiring diagram differ from an electronics schematic diagram?
2. How is a three-conductor cable drawn in an electrical construction wiring diagram?
3. Why should black wires be connected to a single-pole switch in residential wiring?
4. How are switches symbolized in one-line wiring diagrams?
5. What does the symbol of a 5 mm diameter circle in a one-line residential wiring diagram signify?
6. What is the smallest wire size recommended by the National Electrical Code for residential wiring?
7. What name is conventionally used in residential wiring for a SPDT switch?

8. How are 230 volts obtainable from a typical 3-wire residential service entrance box?

9. What are the advantages and disadvantages of remote control residential wiring?

10. Why are arrows used in one-line wiring diagrams in architectural drawings?

11-1
Symbols
Architectural electric wiring is less complicated than that found in most electronic circuits. There are fewer components and the wiring consists mainly of simple parallel (and occasionally series-parallel) circuits. Graphic electrical wiring symbols should conform to those adopted by the Institute of Electrical and Electronic Engineers (IEEE) in their standard 315-71 (ANSI Y32.9 1972).

Most lighting fixture and outlet receptacle symbols are based on a circle. The circle is obviously drawn to represent the round, hexagonal, or rectangular metal box to which the cables or raceways are attached. The ceiling lamp fixture symbol (Fig. 11-1b) is a simple circle. Any departure from the simple fixture requires additions to this symbol such as the letter S when the ceiling fixture has a *pull-chain switch*.

As shown in Figure 11-2, all outlet *receptacle* symbols are also based on the circle. When only one appliance plug is capable of connection to a receptacle, then only a single line crosses the circle symbol. The letter G (Fig. 11-2a and b) shows that the receptacle contains a grounded pin jack. Note the other letter codes at the lower right of the circles for *specialized* receptacle symbols. The range outlet symbol (Fig. 11-2h) contains three horizontal lines; this indicates that three *conductors* are to be connected to the receptacle and *not* three receptacles.

As previously noted the letter S adjacent to a *circle* symbol indicates a lamp fixture having a *switch* attached. All switch symbols in architectural electric schematics are based on the letter S except for panel or entrance switchboards. The three-way switch S_3 (Fig. 11-3b) is actually a single-pole double-throw switch (SPDT); two such switches are often used to control a lamp from two different locations (Fig. 11-8b). The four-way switch S_4 is a DPST switch that is always closed in either position; it is *never* open. It is used to open or close two lines simultaneously (Fig. 11-8b). A simple single-pole switch S is used to open or close only the *hot line* to turn a lamp or appliance OFF or ON.

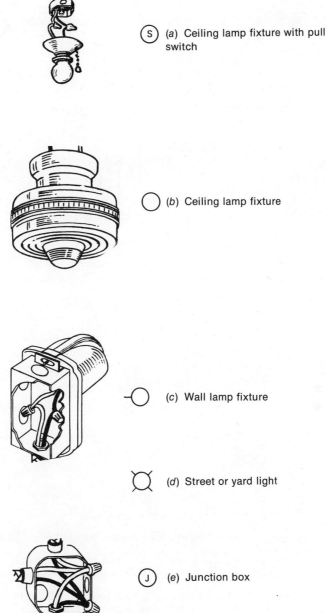

(S) (a) Ceiling lamp fixture with pull switch

(b) Ceiling lamp fixture

(c) Wall lamp fixture

(d) Street or yard light

(J) (e) Junction box

Figure 11-1 Lighting outlets and their symbols.

Most buildings contain at least five completely separate circuits eventually joined near the point where the service lines are attached to the building. The *service entrance switches* (Fig. 11-4*a* and *b*) serve to open or close all circuits or an individual

(a) Single receptacle outlet

(b) Duplex receptacle outlet

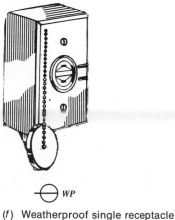

(c) Special purpose receptacle outlet

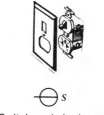

(d) Switch and single receptacle

(e) Duplex receptacle split wired

(f) Weatherproof single receptacle

(g) Single floor receptacle

(h) Range outlet

Figure 11-2 Receptacle outlets and their symbols.

circuit. The pictorials of Figure 11-4*a* and *b* show handles which when pulled will disconnect main circuits; these *disconnect blocks* also contain *cartridge fuses* for protection against short circuits and overloads. Entrance switch boxes are also available containing *circuit breakers* (Fig. 11-3*f*) instead of fuses. Individual low power (15 amperes and 20 amperes) branch circuits are disconnected by unscrewing *plug fuses*.

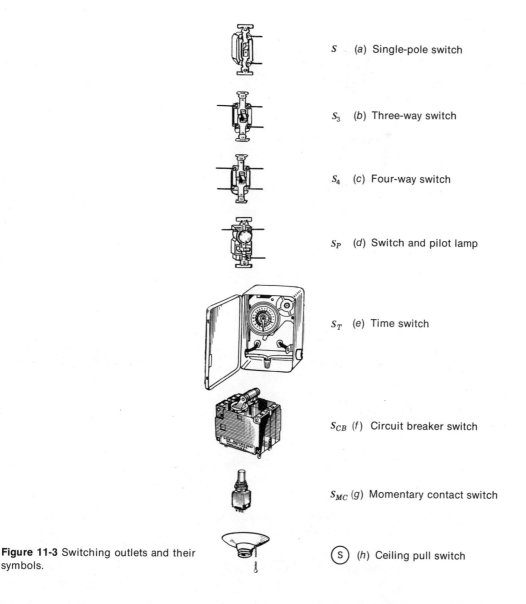

S (a) Single-pole switch

S_3 (b) Three-way switch

S_4 (c) Four-way switch

S_P (d) Switch and pilot lamp

S_T (e) Time switch

S_{CB} (f) Circuit breaker switch

S_{MC} (g) Momentary contact switch

(S)   (h) Ceiling pull switch

Figure 11-3 Switching outlets and their symbols.

Note that all disconnect switch symbols (Fig. 11-4) are *rectangular* in shape. The *darkened* (shaded) symbol indicates a *main entrance* switch, while the clear rectangles symbolize switches wired after the service-entrance switch and are not used to disconnect all service. The symbol for an externally operated disconnect switch (Fig. 11-4c) includes a bent line (to represent the switch handle).

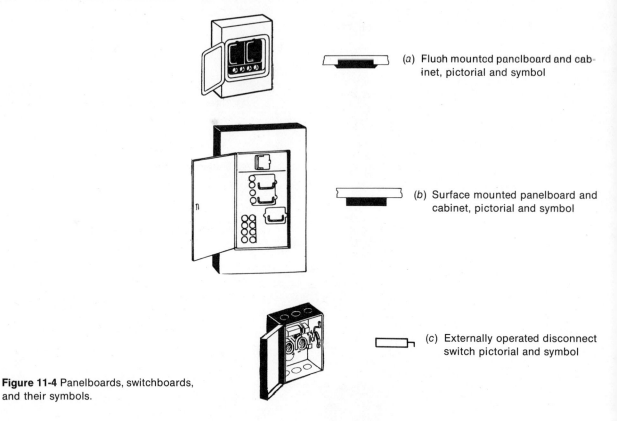

(a) Flush mounted panelboard and cabinet, pictorial and symbol

(b) Surface mounted panelboard and cabinet, pictorial and symbol

(c) Externally operated disconnect switch pictorial and symbol

Figure 11-4 Panelboards, switchboards, and their symbols.

11-2 Single-Line Diagrams

A fundamental prerequisite of any adequate wiring installation is conformity to the safety regulations applicable to the building. The approved *national* standard for electrical safety is the National Electrical Code (NEC). Some states, cities, or counties counties have added further restrictions to the NEC standards. In addition to the NEC standards, the electrical draftsman (as opposed to the electronic draftsman) must become familiar with *architectural* standards and codes.

Since most power conductors are cabled or grouped in *raceways* (conduit) rather than in single conductors, architectural wiring diagrams are drawn with single lines showing conduit or multiconductor cables and *not* for *each* conductor.

Figure 11-5 shows the differences between electronic and architectural electric wiring diagrams. A single line in architectural wiring (Fig. 11-5a) implies *two* conductors in a conduit or cable. *Unbroken solid lines, A,* mean that the cable or raceway is *concealed* in the ceiling or wall. Line *A* therefore shows that a 2-wire cable is concealed in the ceiling or wall between a

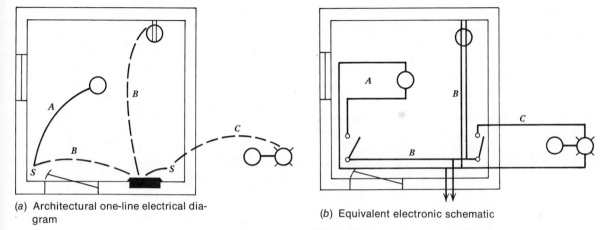

(a) Architectural one-line electrical diagram

(b) Equivalent electronic schematic

Figure 11-5 Comparison of electronic and architectural electric wiring diagrams.

ceiling lamp fixture and a single-pole switch on the wall.

A *long-dashed line* is shown between switch *S* and the entrance switch box in Figure 11-5*a*. The code specifies that this 2-conductor cable is *concealed* in the *floor*. Similarly, the 2-conductor cable *B* connecting the duplex receptacle to the entrance switch box is also concealed in the floor. Note that the electronic equivalent diagram does not indicate *where* the cable is to be placed.

A third circuit *C* (Fig. 11-5*a*), drawn with a *short-dashed line* from the building to a yard light on a pole, indicates that the 2-wire cable is exposed (or strung overhead) from the building to the pole. Solid or dashed lines are *not* intended to show actual paths of cables between outlets and switches but merely indicate schematically which outlets are controlled by the switches. According to code, branch circuit wiring locations are the responsibility of the experienced journeyman electrician. The equivalent electronic schematic for Figure 11-5*a* is shown in Figure 11-5*b*.

A complete wiring diagram of a kitchen is shown in Figure 11-6. Recall that all wiring lines are 2-conductor unless otherwise indicated. *Three* short diagonal lines are added to the cable line when *three* conductors are cabled, as shown connected to the recessed fluorescent lamp fixture, *R* (Fig. 11-6). This ceiling fixture (*R* for recessed) is controlled by two 3-way switches S_3 from either door; three-way switches require 3-wire cable (Figs. 11-3*b* and 11-8*b*).

The direction arrow marked 4 (Fig. 11-6) is a "*home run*" for

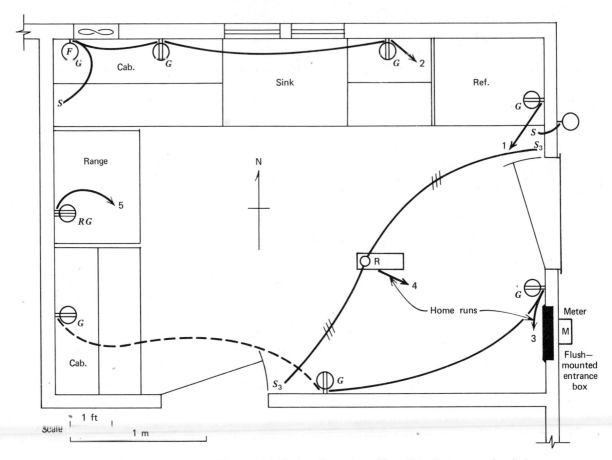

Figure 11-6 Kitchen floor plan with outlets, fixtures, and switches.

the ceiling fixture. Any line with an arrow and number represents a numbered branch circuit home run to a panelboard or service entrance switch. It means that power for the ceiling fixture is obtained from terminal 4 of the service entrance switch located in the lower right-hand corner of Figure 11-6. Five home runs (branch circuits 1, 2, 3, 4, and 5) are shown in Figure 11-6.

The remaining circuits in the kitchen plan (Fig. 11-6) consist of simple 2-wire cable circuits because of the lack of short diagonal lines crossing the cables. Three duplex receptacle outlets are wired in parallel along the east, south, and west walls of the room to form circuit 3. Circuit 1 supplies power to a refrigerator and the outside lamp fixture; the lamp is controlled through an interior switch S near the door. The switch-controlled exhaust fan F and two duplex receptacle outlets comprise branch circuit 2.

The electric range receptacle outlet (RG) is always independently connected with a heavy 3-wire cable to a heavily fused disconnect block or circuit breaker in the service entrance box as shown in Figures 11-6 and 11-8. This circuit (5) supplies double (230 volts) the voltage required by the lighting load (115 volts).

As noted in Section 11-1, the letter G near the outlets indicates the presence of a common ground terminal; an added requirement of the NEC demands that a common ground wire is connected to every outlet. Two-conductor *nonmetallic sheathed cable* now includes a bare ground wire for this purpose although diagonal crossing lines are not added to the cable line. In larger installations such as commercial and industrial buildings, *metal conduit* is required. Where single-phase 2-wire and dc circuits are installed, the conduit itself serves as the ground conductor. In 3-wire single-phase and 4-wire three-phase circuits, a neutral conductor is used to carry any current imbalance.

11-3 Floor Plan Layouts Proper illumination is an essential element of modern living. The amount, type, and location of illumination required should be fitted to the various tasks carried on in the home, office, or factory. Good lighting therefore requires careful planning by the architect in locating lighting fixtures, switches, and convenience outlets.

All rooms should have either a ceiling fixture, wall fixtures, or at least *one* convenience outlet on one wall connected to a portable lamp and controlled by a wall switch located near the doorway. Each closet should have a lighting outlet controlled either by a wall switch near the door or an automatic door-type switch; lighting fixtures with pull-chain switches (Fig. 11-1a) are still being used but are not recommended.

Convenience outlets preferably should be located near the ends of a wall space rather than near the center, thus reducing the likelihood of being concealed behind large furniture. One receptable outlet should be provided for each 1.22 m (4 lin ft) of kitchen work surface frontage for connection of portable appliances and located approximately 1.12 m (44 in.) above the floor. In rooms other than the kitchen, convenience outlets should be placed approximately 30 cm (12 in.) above the floor and so that no point along the floor line in any usable wall space is more than 1.83 m (6 ft) from an outlet. One outlet is needed for each 4.6 lin m (15 ft) of hallway or if it has over 2.3 m² (25 ft²) of floor area. Indoor or outdoor waterproof outlets (Fig. 11-2*f*) should be located 46 cm (18 in.) above grade and be controlled by a wall switch inside the building.

Wall switches are symbolized with the letter *S* and a number or letter representing the type (Fig. 11-3). Wall switches should normally be located at the *latch-side* of doors or at the traffic side of arches and within the room or area to which the control is applicable. Wall switches are normally mounted approximately 1.22 m (48 in.) above the floor. Hallways, stairways, and rooms with more than one entrance 3.0 m (10 ft) or more apart should be controlled with a switch at each entrance (Fig. 11-6).

**11-4
Convenience
Outlet
Details** The duplex receptable outlet circuit (Fig. 11-8) shows a single outlet connected to the service entrance box and a parallel cable connection to another outlet (not shown). Almost all 115 volt convenience outlets are connected in parallel with all *black* wires attached to the *brass colored* screws and all *white* (ground) wires to the *silver-colored* screws. The *green-colored* hex-head screws are for the continuation of the common ground wire. Several convenience outlets may be connected in series-parallel through a wall switch.

**11-5
Switching
Circuit
Details** The simple series circuit controlling a ceiling lamp is shown in Figure 11-7*a*; a schematic diagram is shown instead of the one-line diagrams of Figure 11-6. A switch must always be capable of disconnecting the *hot* (black) line; therefore the black wires (B) are connected to the switch and the center contact of the lamp receptacle. The white (W) ground wire should be connected to the screw shell of a lamp receptacle.

In the common residence circuit (Fig. 11-7*b*), a ceiling fixture is controlled by either of two 3-way switches or by the 4-way switch. It is necessary to use 3-wire cable (Figs. 11-6 and

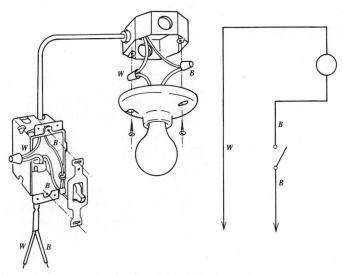

(a) Lamp circuit controlled with single-pole switch

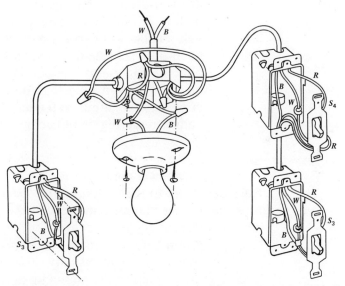

(b) Lamp circuit controlled from either of two 3-way switches and one 4-way switch

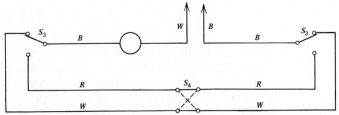

Figure 11-7 Pictorial and schematic lamp control circuits.

(c) Schematic of above

11-7*b*). The red (R) and white (W) wires are called *travelers* and are connected to the *brass-colored* switch terminals. The *black* or *copper-colored hinge-point* terminal should always be connected to the black (B) hot line. Note in the schematic of Figure 11-7*b* that *either* 3-way switch can turn the lamp OFF or ON. The 4-way switch is always closed in either position; it is never open. It serves to reverse the red and white travelers and in so doing will also control the lamp.

To facilitate switch maintenance and reduce wiring cost, a *low-voltage* remote switching system is becoming popular. In a *remote-control system,* relays that perform the actual switching of the current are located in each outlet. The relays are controlled by small switches in a 24 volt circuit supplied from an isolating 115 or 230 volt *step-down transformer.* One advantage of the system is that the control wiring uses less expensive 16 or 18 AWG insulated wire instead of heavier power cable. Conventional power cable is only connected to the lighting or appliance outlets. Remote switches placed at the entrance to the building can control distant lights, exhaust fans, heating units, and the like. Since these remote switches carry low voltage and current, their life is extended and there is less danger to personnel repairing or replacing them.

11-6 Service Entrance Details The power company usually connects the service wires from the street to the building. The service may be carried to the building via underground service entrance or overhead service drop, depending on the location of the distribution lines (underground or overhead). In applying for electrical service, the power company usually advises the customer of any restrictions, such as the minimum height and location of the weatherproof *entrance head* and *pull-off* (Fig. 11-8). The power company only furnishes the meter socket and the cable from the street to the building.

The service-entrance switch (Fig. 11-8) is usually installed *within* the building and provides both overcurrent protection and disconnecting means. The two insulated *entrance cable* wires are connected to the upper left fused disconnect block (Fig. 11-8). This block contains the main *cartridge fuses* and also serves to disconnect the 230 volt service when pulled from the box. The third (ground) conductor is *not* insulated and

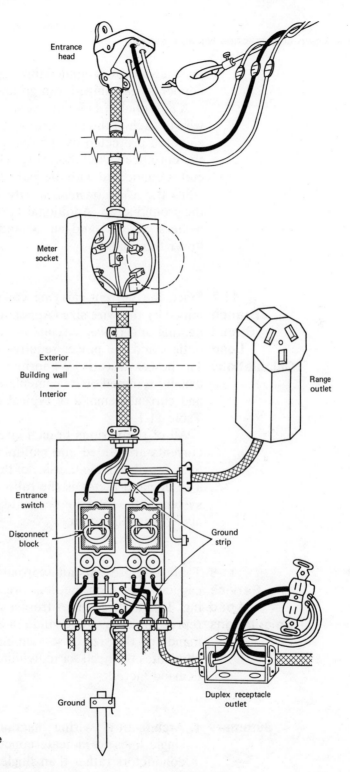

Entrance
head

Meter
socket

Exterior

Building wall

Interior

Range
outlet

Entrance
switch

Disconnect
block

Ground
strip

Duplex receptacle
outlet

Ground

Figure 11-8 Typical residential service
entrance.

must always be grounded through the switch box to an external copper or galvanized iron ground rod (Fig. 11-8).

A potential of 115 volts exists between this *neutral* or ground wire to each of the two insulated wires. The 115 volt branch circuits are protected by the four *plug fuses* in the lower part of the entrance switch box (Fig. 11-8). Each 115 volt branch circuit is connected with the *black* or *hot wire* to the fuse terminal while the *white* or *neutral wire* is grounded within the box to the ground strap. Additional 115–230 volt service for hot water heaters, dryers, and air conditioners is obtainable from the lower terminals.

**11-7
Branch
Circuit
Load
Calculations**
Since the current-carrying capacity of copper wire is determined by the wire size (Appendix B), the wire gauge size of all lighting and power circuits is determined from the number of outlets and the power requirements of connected appliances. The power in watts (W) normally consumed by an appliance is usually stamped on its nameplate. Average values of the power and current demand of typical electrical devices are listed in Table 11-1.

All of the various branch circuit, appliance, and receptacle currents are added and multiplied by an appropriate *demand factor* to determine the size of the service entrance conductors. The demand factor is the ratio of the maximum demand of a system to the total connected load on the system. This factor is always less than unity.

**11-8
Service
Load
Calculations**
Table 11-2 presents an approximation of the minimum service capacity based on the floor area of a residence or small building. This table emerges from a study of many installations and represents an average guide in estimating both the current demand and the size of service entrance conductors. Table 11-2 eliminates the need for individual branch circuit calculation and demand factors.

Summary
1. Architectural wiring diagrams are simplified by drawing single lines to indicate conductors in a raceway or cabled conductors rather than single conductors.

Table 11-1
Average Current and Power Drawn by Domestic Loads

	Power (watts)	Current (amperes)
Air conditioner, room, 3/4 ton	1,200	10.4
Blanket, electric	170	1.5
Dryer, clothes (230 volt)	5,000	21.7
Freezer	350	3.0
Furnace, hot air, oil fired	800	7.0
Heater, water standard (230 volt)	2,500	10.9
Iron	1,000	8.7
Lamp, table	150	1.3
Mixer, food	150	1.3
Pump, water	700	6.1
Radio/record player	110	1.0
Range (230 volt)	10,000	43.5
Refrigerator	320	2.8
Sewing machine	75	0.6
Television, color	330	2.9
Toaster, automatic	1,100	9.6
Vacuum cleaner	630	5.5
Washer, dish	1,200	10.4
Washing machine, automatic	510	4.4
Waste disposer	440	3.8

Table 11-2
Approximate Service Entrance Feeder Size Based on Total Floor Area for a Given Occupancy

Floor area		Minimum Service (amperes)	Feeder Size (AWG No.)
Square Meters	Square Feet		
Up to 93	1000	125	2
94–186	1001–2000	150	1
187–279	2001–3000	200	000

2. Fixture symbols usually consist of coded circles.

3. Circles are also basically used to represent convenience outlets (but crossing lines are included).

4. Switches are symbolized simply with the letter *S* and a number or letter indicating the type of switch.

5. All construction wiring must adhere to the National Electrical Code (NEC).

6. Convenience receptacles are wired in parallel.

7. Some lighting fixtures and receptacles are controlled by series-connected wall switches.

8. The wire size of conductors is determined by the *maximum* connected current load.

Problems **11-1** Find the number of outlets in branch circuit 1 of the floor plan shown in Figure 11-6.

11-2 How many electrical outlets are in branch circuit 2 of the floor plan in Figure 11-6.

11-3 Draw a one-line schematic of a ceiling fixture controlled by a wall switch, similar to Figure 11-7a, except that the supply voltage is fed into the octagonal box instead of the switch box.

11-4 Draw a one-line schematic of a ceiling fixture controlled by two 3-way wall switches, similar to Figure 11-7b, except that the supply voltage is fed into the right-hand 3-way switch box.

11-5 Determine the branch wire size needed if *all* convenience outlets in the floor plan of Figure 11-6 are connected to the *same* branch circuit. Assume that a toaster, food mixer, radio, iron, and room air conditioner are connected to the outlets. (Exclude range and refrigerator outlets.)

11-6 a. How much total power is consumed if *all* devices in problem 11-5 (including range and refrigerator) in the floor plan of Figure 11-6 are connected at the same time.

   b. Is it a valid assumption that all loads are operating simultaneously? Explain.

11-7 Measure and draw the floor plan of your classroom to a convenient scale. Add all outlets and the one-line wiring diagram, assuming that all outlets and fixtures are on a single branch circuit. Calculate the total and minimum wire size required using a 115/230 volt 3-wire service.

11-8 Plan and draw the wiring diagram to a convenient scale for the first floor plan shown in Figure 11-9.

11-9 Plan and draw the wiring diagram to a convenient scale for the residence shown in Figure 11-10.

Figure 11-9 First floor plan of residence for Problem 11-8.

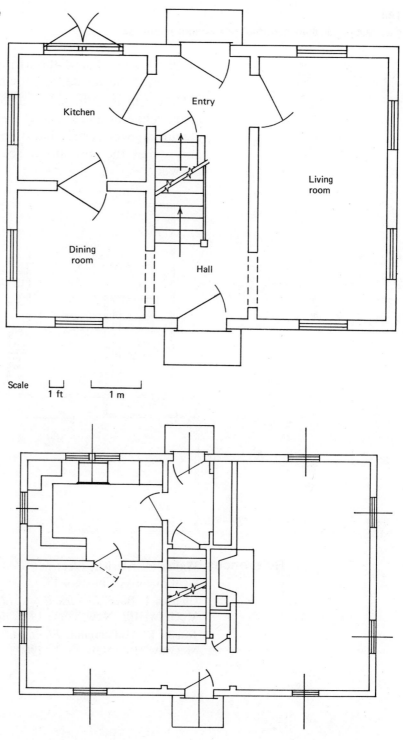

Kitchen

Entry

Living
room

Dining
room

Hall

Scale
1 ft   1 m

Figure 11-10 Partial floor plan of residence for Problem 11-9.

Scale
1 ft   1 m

11-10 Plan and draw the wiring diagram to a convenient scale for the residence shown in Figure 11-11.

Now return to the self-evaluation questions at the beginning of this chapter and see how well you can answer them. If you cannot answer certain questions, place a check next to them, and review the appropriate sections of the chapter to find the answers.

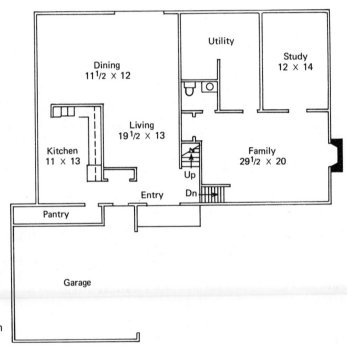

Figure 11-11 Floor plan for Problem 11-10.

References Walter N. Alerich, *Electrical Construction Wiring*, American Technical Society, Chicago, 1971.

Charles J. Baer, *Electrical and Electronics Drawing*, Third Edition, McGraw-Hill, New York, 1973, pp. 267–286.

Joseph F. McPartland, *Electrical Systems Design*, Second Edition, McGraw-Hill, New York, 1960.

Chapter 12 Industrial Control Wiring Diagrams

Instructional Objectives

1. To become aware of the differences among industrial power and residential wiring diagrams compared to electronic wiring diagrams.
2. To learn to read industrial control wiring diagrams.
3. To develop familiarity with electrical symbols used in power distribution diagrams.
4. To learn some basic differences between ac and dc motors.
5. To learn why special starting circuits are required for industrial motors.
6. To learn some operations of electrical protective devices and their circuits.

Self-Evaluation Questions

Test your prior knowledge of the information in this chapter by answering the following questions. Watch for the answers as you read the chapter. Your final evaluation of whether you understand the material is measured by your ability to answer these questions. When you have completed this chapter, return to this section and answer the questions again.

1. How does the symbol for a fuse in an industrial control diagram differ from that in an electronic schematic?
2. What is a "ladder" diagram?
3. How does a ladder diagram differ from a residential wiring diagram?
4. How does a power contact symbol differ from the symbol for a capacitor?
5. How may the drawing of a three-phase power distribution system circuit diagram be simplified?
6. For what general category is a circle symbol used in industrial control diagrams?
7. Why must a rectangular-shaped symbol always include an identification code letter or number?
8. In what physical position are the input power lines drawn in a conventional industrial control wiring diagram?

9. How does the wye transformer connection symbol differ from the delta connection symbol?
10. In what part of an industrial control diagram are contactor and relay coils drawn?

**12-1
Components and
Symbols**

Industrial control wiring diagrams are considerably different from schematic diagrams for electronic communication in various ways. Electromechanical components, such as switches, relays, solenoids, and push buttons, are extensively used. Some of these control diagrams also include mechanical linkages, cams, or pneumatic devices. Finally, many industrial control circuit symbols have a different meaning than those of similar symbol shape that we have used up to now.

The *circle* is usually represented as an outlet in architectural diagrams. In industrial control diagrams, the circle may represent a relay coil (Fig. 12-4a) or a motor (Fig. 12-5e). On the other hand, some industrial control symbols are similar to those used in electronic schematics: capacitors, inductors, transformers, momentary or push button switches, and so forth.

Note that resistors are usually represented by a rectangle in industrial wiring diagrams instead of the customary zigzag line (Fig. 12-1). Partly because a rectangle symbol may sometimes be used to represent a relay coil device designations are placed within or adjacent to the resistor rectangle. The designation may be a combination of consecutive numbers and assigned letters. The numbers, which differentiate between functions or components of the same class, *precede* the letter designations, such as *3R*. Standard device designations are to be found in Appendix A. Note in Figure 12-1b and c that a tapped or adjustable resistor may also be symbolized in a manner similar to that used in electronic symbols.

Industrial control switches usually must be capable of making or breaking circuits that carry a much heavier current than do electronic circuits. Furthermore, since much higher voltages are used in industrial power circuits, arcing is apt to occur when a switch is opened. For these reasons, industrial power switches are physically much heavier and larger.

Switch symbols are almost the same as those used in electronic circuits (Fig. 12-2). The harmful effect of an electric arc,

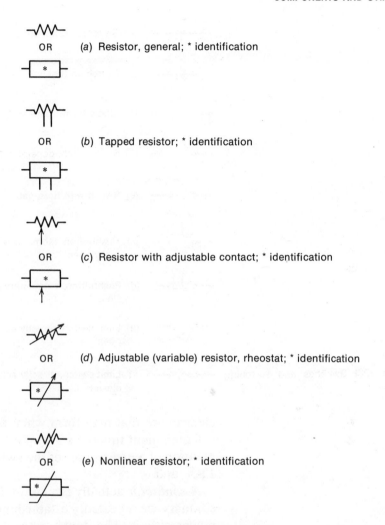

Figure 12-1 Resistor symbols.

which forms when any highly inductive circuit is interrupted, is somewhat reduced by a switch having a *horn gap* (Fig. 12-2e). Device designations should also identify all switches.

The electronic symbol for a fuse is usually replaced with a rectangle (Fig. 12-3). Note that the low resistance of a fuse may be indicated by drawing the circuit line lengthwise through the rectangle to differentiate this symbol from that of a resistor or relay coil. Another general form for the fuse symbol resembles a cartridge fuse with end caps (Fig. 12-3a).

The double rectangle indicates that the fuse is contained in a chamber filled with oil. The special oil serves to extinguish the

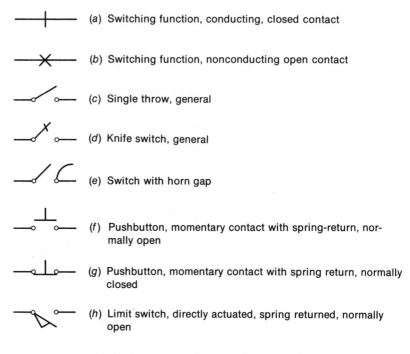

(a) Switching function, conducting, closed contact

(b) Switching function, nonconducting open contact

(c) Single throw, general

(d) Knife switch, general

(e) Switch with horn gap

(f) Pushbutton, momentary contact with spring-return, normally open

(g) Pushbutton, momentary contact with spring return, normally closed

(h) Limit switch, directly actuated, spring returned, normally open

(i) Limit switch, directly actuated, spring returned, normally closed

Figure 12-2 Switches and switching functions.

electric arc that may form when the fuse blows.

Component functions are often combined, such as the combination of a fuse as part of the switch lever, as shown in Figure 12-3c and e.

A *contactor* actually is a heavy-duty relay; as such, its heavy contacts are repeatedly establishing and interrupting an electric power circuit. The contactor coil is usually symbolized by a circle containing a code letter and/or a number for symbol identification. Symbol identification is always necessary. In industrial schematic diagrams, the coil is often not drawn adjacent to its contacts because the coil and contacts are usually in *different* circuits.

If closed contacts are opened when the relay coil is energized, the contacts are considered *normally closed* (n.c.). An n.c. contact is represented by a diagonal line drawn across the two parallel lines of the contact symbol (Fig. 12-4g). Conversely, a *normally open* (n.o.) contact symbol does *not* contain

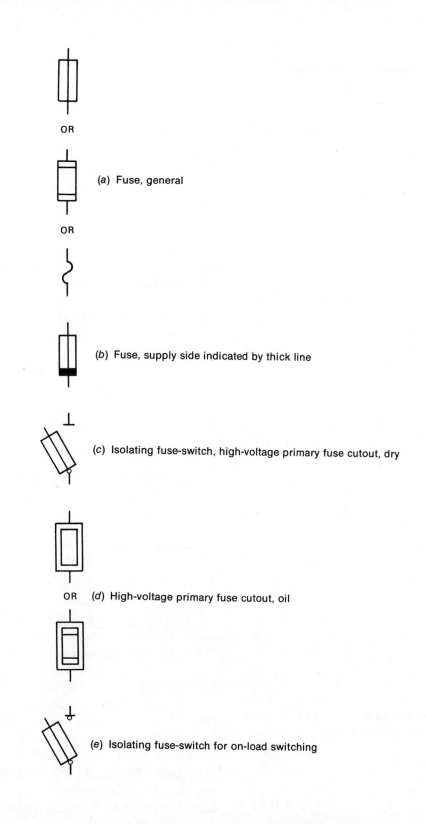

OR

(a) Fuse, general

OR

(b) Fuse, supply side indicated by thick line

(c) Isolating fuse-switch, high-voltage primary fuse cutout, dry

OR   (d) High-voltage primary fuse cutout, oil

(e) Isolating fuse-switch for on-load switching

Figure 12-3 Fuse symbols.

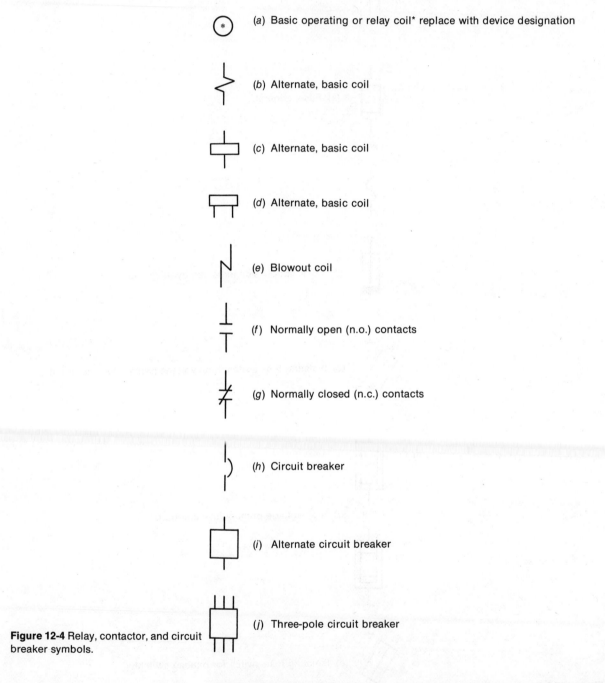

(a) Basic operating or relay coil* replace with device designation

(b) Alternate, basic coil

(c) Alternate, basic coil

(d) Alternate, basic coil

(e) Blowout coil

(f) Normally open (n.o.) contacts

(g) Normally closed (n.c.) contacts

(h) Circuit breaker

(i) Alternate circuit breaker

(j) Three-pole circuit breaker

Figure 12-4 Relay, contactor, and circuit breaker symbols.

the diagonal line (Fig. 12-4f). All contacts are shown in their de-energized state—always!

Care should be taken when drawing the symbol for contacts,

since it may look like a capacitor unless the two vertical or horizontal lines are sufficiently separated and parallel (Fig. 12-4*f*).

A *circuit breaker* (CB), unlike a relay, senses and clears the current flow in the *same* line in the event of an overload. Consequently, in a CB, the contacts must handle the same current as the trip coil. But in a contactor or relay, the relay coil current is usually lower (sometimes higher) than the current carried by the contacts; the contacts are usually in separate (higher power) circuits. *Overload* (OL) *relays,* for example, sense high current in the power circuit but trip low current contacts in the control circuit, as distinguished from fuses and overload circuit breakers (OCB), which sense and clear high currents.

A *blowout coil,* shown symbolically in Figure 12-4*e,* usually is a part of an *air circuit breaker*. Current through the coil creates a magnetic field that serves to extinguish the arc formed when the contacts are opened.

The circle is used as the basic shape for rotating electric machinery symbols (Fig. 12-5*a*). As noted, symbol identification by letter and number must be placed either within or adjacent to the circle.

All electric motors and generators require a magnetic field for operation. The magnetic field in dc motors and some ac motors is usually provided by a winding on the stationary pole pieces; the field winding is therefore symbolized as an inductor. As shown in Figure 12-5*f,* the inductor symbol is drawn in the same line as the *armature* or *rotor* when the machine has a *series field*. A *shunt field* inductor symbol (Fig. 12-5*g*) is drawn *in shunt* or parallel with the armature or rotor. When both a series and shunt field is used, it is known as a *compound field* winding.

Either the complete or incomplete helical inductor symbol may be used to represent a transformer winding (Fig. 12-6*a*). A *three-phase* transformer usually contains six separate windings (Figs. 12-6*b* and 12-13*a*), one pair of *primary* and *secondary* windings (see Section 4-2) for each phase.

As noted in the last chapter, architectural wiring diagrams are simplified by using a single line to indicate a multiconductor cable. Since three-phase ac circuits may require three conductors and ground, industrial control diagrams often use the simplified form of three-phase transformer which requires only two inductor symbols to be drawn, as shown in the left portion

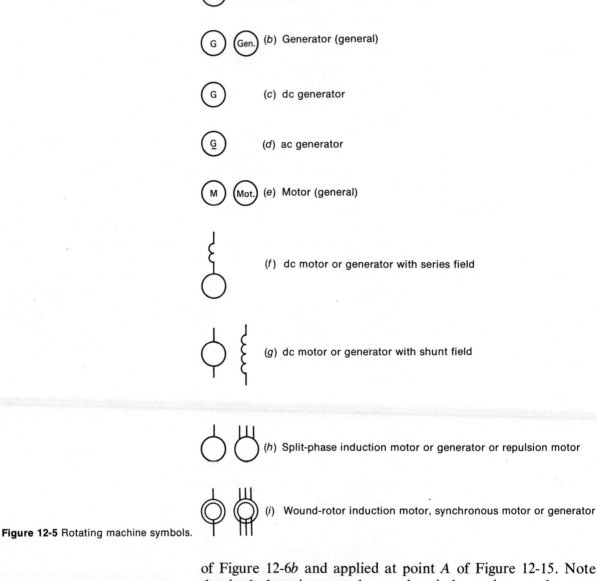

(a) Basic

(b) Generator (general)

(c) dc generator

(d) ac generator

(e) Motor (general)

(f) dc motor or generator with series field

(g) dc motor or generator with shunt field

(h) Split-phase induction motor or generator or repulsion motor

(i) Wound-rotor induction motor, synchronous motor or generator

Figure 12-5 Rotating machine symbols.

of Figure 12-6*b* and applied at point *A* of Figure 12-15. Note that both the primary and secondary inductor is tapped to represent the input and output three-phase line.

Three-phase windings may be variously interconnected; it is therefore important that *wye* or *delta connection* symbols (Fig. 12-13*a*) accompany the transformer symbol when a simplified one-line diagram is used for a three-phase circuit. The connec-

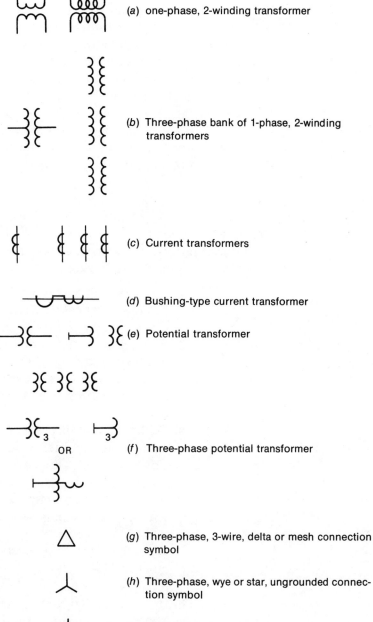

(a) one-phase, 2-winding transformer

(b) Three-phase bank of 1-phase, 2-winding transformers

(c) Current transformers

(d) Bushing-type current transformer

(e) Potential transformer

(f) Three-phase potential transformer

OR

(g) Three-phase, 3-wire, delta or mesh connection symbol

(h) Three-phase, wye or star, ungrounded connection symbol

(i) Three-phase, wye or star, grounded connection symbol

Figure 12-6 Transformers and their connection symbols.

tion symbols are also shown in Figure 12-6*g, h,* and *i,* and applied at point *A* in Figure 12-15.

A *current transformer* is an instrument transformer used to connect ammeters and current coils of instruments into high-current lines. The cable in which current is to be measured can pass directly through a hole in the transformer assembly; the cable thereby becomes the primary. The inductor symbol is therefore drawn *on the line,* as shown in Figure 12-6*c* (also see point *B* in Fig. 12-15). Similarly, *potential transformers* are used to connect a voltmeter to the low voltage side to measure potential of a high voltage ac circuit (Fig. 12-6*e*).

12-2 Schematic Layout Conventions As previously noted, conventional signal flow in electronic schematic diagrams is from left to right. The elementary diagram for industrial control and power distribution is also drawn using the following similar conventions adopted by industry.

1. Input power begins at the top and progresses downward between vertical *feed lines* to the motor load; this is sometimes called a *ladder diagram* (Fig. 12-11).
2. When a dc circuit is drawn, the left-hand vertical feed line is considered as *positive* (ungrounded).
3. Three vertical lines are drawn for a three-phase ac diagram or may be further simplified by drawing a single vertical line.
4. Switches, circuit breakers, and fuses are generally drawn near the top and left side of the diagram.
5. Contactor and relay coils are usually drawn near the negative or right-hand dc feed line.
6. It is common practice to include reference designations within the symbols for resistors, contactors, and relay coils.
7. When horizontal and vertical lines are intended to make an electrical connection, the connection is *always* indicated by a heavy dot *where the lines cross.*
8. If it is intended that a crossover of lines is *not* electrically connected, the dot should be omitted.
9. Each connection line between devices is usually assigned a reference number or letter, generally beginning at the upper left of a ladder diagram.

10. The horizontal line having the most symbols usually determines the spacing between vertical feed lines.
11. Balance of components within the diagram is desirable for easier circuit tracing.
12. Often the power circuit is drawn with heavier or bolder lines than the control circuit (which is shown in thinner lines) for easier location and identification.

12-3
AC Motor
Control
Circuits

The motor control circuit is a device, or group of devices, that governs the starting, protection, running, speed regulation, and stopping of a motor. The ladder diagram (Fig. 12-7) provides a convenient system for identifying components and functions of components in these complex circuits.

If full line voltage is connected directly to a motor, the starting current is several times the normal running current of a

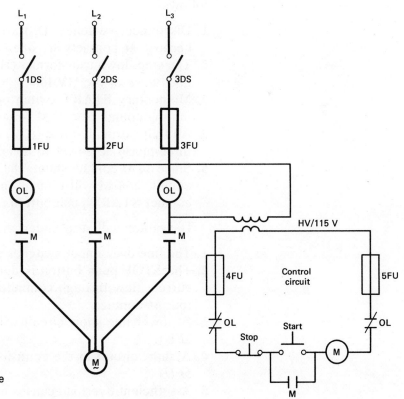

Figure 12-7 Automatic across-the-line ac starter circuit.

motor. Large ac motors draw excessive currents when started *across the line* at full voltage, occasionally causing severe supply line disturbances. The low-power automatic across-the-line ac starter circuit (Fig. 12-7) for smaller motors contains line fuses and overload contactors for protection against the effects of excessive starting and overload currents. Smaller motors may be started across-the-line if such starting does not cause serious line voltage reduction.

Figure 12-7 shows how the *power circuit* is drawn near the top of the diagram and connected across the high-voltage (HV) three-phase ac line. Note also that the *control circuit* is drawn as a lower step in this simple ladder diagram. The control circuit is isolated from the power circuit through a step-down transformer to 115 volts START–STOP switches and contactor coil (M) are wired in this separate lower voltage and current control circuit.

Operation of the ac motor starter (Fig. 12-7) proceeds as follows.

1. Disconnect switches (DS) are closed. Motor will not start because *M* contacts are open.
2. The step-down transformer (HV/115 volts) is energized from one *phase* of the HV line.
3. Momentary START switch push button is depressed, permitting contactor coil M to be energized.
4. All four normally open (n.o.) *M* contacts close, causing full three-phase ac power to be applied to the motor.
5. Since an *M* contact shunts the START switch as an *interlock* circuit, coil M will remain energized when the momentary contact START push button is released.

The motor will stop under any of the following conditions.

1. The line-disconnect switches (DS) are opened.
2. The STOP push button is depressed. This action will stop current flow through contactor coil M, thereby opening all four *M* contacts.
3. A power line short circuit will open line fuses (1*FU*, 2*FU*, 3*FU*).
4. A short circuit in the control circuit will open fuses (4*FU*, 5*FU*).
5. A sufficient overload causes normally closed (n.c.) OL contacts in the control circuit to open. When the control circuit

is opened, then contactor coil M is de-energized, and the line contacts (M) are opened.

6. A temporary reduction or loss of line voltage may be sufficient also to de-energize coil M, requiring restarting the motor when voltage is restored.

Several starting circuits have been devised for applying *reduced* ac voltage to large *polyphase* motors. When reduced voltage for starting is used, excessive starting currents are also reduced. The *primary resistor type starter* shown in Figures 12-8a and 12-9 reduces the line voltage by connecting a resistor in each of the three lines. Primary resistor-type starters are ideally suited to applications such as conveyors, textile machines, or other delicate machinery where reduction of starting torque is not of prime consideration.

As discussed in Section 12-3, three-phase windings of an ac motor may be connected wye or delta. The wye connection places two windings in series, while the delta connection places a single winding across the line. The stator windings of a special type of three-phase motor, using this type of starter, require that all six terminals of the three phases be available (Fig. 12-8b). The six leads of the stator winding are switched to a wye connection for starting and a delta connection for normal operation. When wye connected, approximately 58 percent of full line voltage is applied to each winding. Thus the motor is started at reduced voltage, resulting in reduced starting current and less line disturbance.

Wye–delta type starters (Fig. 12-10) are used extensively in large industrial air conditioning installations because they are particularly applicable to starting large motors driving high inertia loads requiring long acceleration times.

12-4 DC Motor Control Circuits

Although ac motors find much wider and general use than dc motors throughout industry, dc motors are better suited to applications requiring control of motor speed. The speed of dc motors varies widely and is mainly determined by the method of adjusting the magnetic field and/or armature voltage. Shunt and compound wound dc motors (Fig. 12-5) are well adapted to speed adjustment from zero up to rated speed and even higher.

Field control is a method of dc motor speed control to obtain speeds *above* a basic speed. The field rheostat 3R in the dc

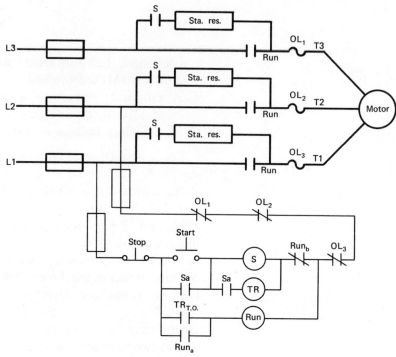

(a) Primary resistor type motor starter

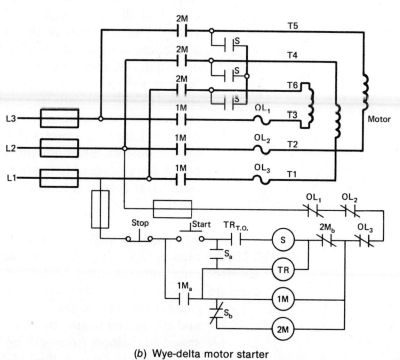

Figure 12-8 Three-phase motor starter circuits. (Courtesy of Westinghouse Electric Corp.)

(b) Wye-delta motor starter

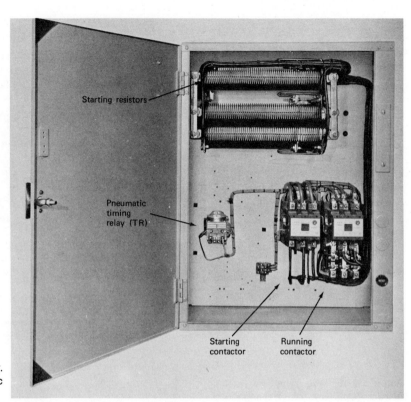

Starting resistors

Pneumatic
timing
relay (TR)

Starting
contactor

Running
contactor

Figure 12-9 Primary resistor-type starter.
(Courtesy of Westinghouse Electric
Corp.)

shunt-motor starter of Figure 12-11 will control the amount of
current flowing through the shunt field winding. Speed control
is achieved by weakening the magnetic field within limits. If the
field is weakened considerably or the *field circuit* is opened,
dangerously high speeds are produced. If the field winding
opens, the *field-loss* (FL) contactor is de-energized and the FL
contacts in the *control circuit* will open and stop the motor.

Operation of the dc shunt-motor (Fig. 12-11) proceeds as
follows.

1. Disconnect switches (DS) are closed, applying full voltage
   across the field circuit.
2. Contact FL in the control circuit closes because the field-
   loss relay (FL) is energized.
3. When the momentary contact START button is depressed,
   contactor coil M in the control circuit is energized.
4. Two *M* contacts in the *armature circuit* close starting the
   motor, and two *M* contacts in the *control circuit* also close.

Figure 12-10 Wye–delta-type motor starter. (Courtesy of Westinghouse Electric Corp.)

5. Auxiliary n.o. contact *M* across the START push button closes and permits the operator to release the push button.
6. When the n.o. *M* contact in the 1*A* dashpot time-delay relay circuit closes, the 1*A* contact in the armature circuit also closes after a brief time delay. One third of the *starting resistance* (1*R*) is shorted out, causing more voltage to be applied across the armature and raising speed.
7. Energizing the 1*A* relay also closes the 1*A* contact in the control circuit, causing relay 2*A* to be energized.
8. After a similar time delay, the 2*A* contacts are closed, and another third of the starting resistance (1*R*) is shorted. Each time that additional resistance is cut out in series with the armature, the armature develops more current and torque and accelerates to a higher speed.
9. Similarly, the 3*A* relay is energized, causing the remainder of the armature's series resistance to be shorted out. The full voltage is then applied to the armature, causing the motor to

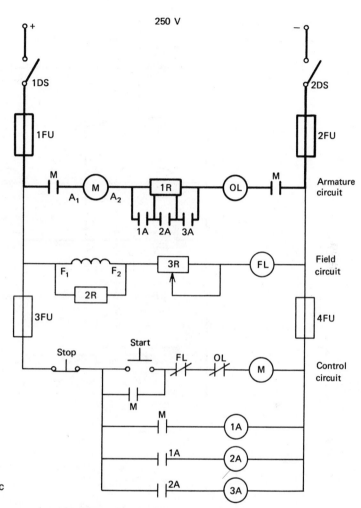

Figure 12-11 Starter circuit for a dc shunt-motor.

operate at "full speed," depending on the setting of the field rheostat (*3R*) and the amount of load.

Since dc electrical power is not usually available and since speed control of ac motors is complicated and expensive, the readily speed-controlled dc motor may be operated on three-phase ac through the use of silicon-controlled rectifiers (SCR). A partial diagram of dc motor speed control from a three-phase supply is shown in Figure 12-12. The six SCRs convert the three-phase ac to a variable dc voltage. A change in the positive trigger voltage applied to the gates G_A, G_B, and G_C changes the

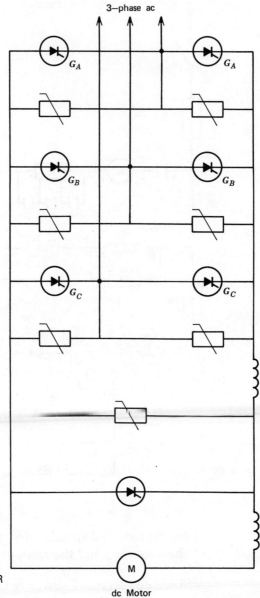

Figure 12-12 Partial diagram of an SCR
dc motor control.

conduction period and therefore the amount of dc voltage output. The gate control circuit is not shown in Figure 12-12.

12-5 Power Distribution Systems

Efficiency of electrical power transmission is improved through the use of higher voltages and lower currents. The use of lower currents results in smaller wire size which reduces installation costs and transmission losses. However, it is difficult and dangerous to consumers to operate motors and other appliances for use at these high transmitted voltages (from 200–600kV). Therefore, for distribution to consumers, ac voltages must be reduced (and currents increased) through the use of *step-down* transformers.

A transformer consists of a primary and one or more secondary windings on a laminated iron core. In a step-down transformer, high voltages connected across the primary winding produce reduced voltages across the secondary winding in the same ratio as the turns ratio of the primary to the secondary.

Three identical transformers may be connected in delta or wye connections (Fig. 12-13a) for three-phase power distribution. Note that the transmitted high voltage (2400 volts) is reduced by the transformation to 120 volts for the single phase lighting loads and three-phase 208 volts for motor loads.

The principal advantage in the use of higher secondary voltage in buildings results from the lower current, which, in turn, reduces the voltage drops across the power lines. Fewer or smaller conductors can be used to transmit the power from the service entrance point to panelboards or other final distribution points. On the other hand, arcs are more likely to occur on protective devices at higher voltages; note the necessity of oil and air circuit breakers at the high voltage side of the circuit and components marked *A* (Fig. 12-14).

A typical power distribution system one-line diagram is shown in Figure 12-14. It also shows the method used to localize an overcurrent disturbance. The protective device closest to the cause of the excessive current should have the first chance to operate. Beginning at a *motor control center (B)* at the bottom of Figure 12-14, it will be noted that five circuit breakers are in the line to the top *utility supply*. In such a ladder diagram, each preceding protective device should be capable,

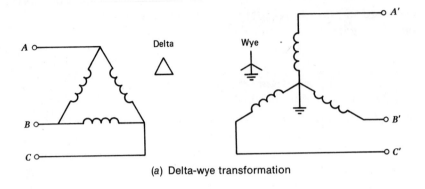

(a) Delta-wye transformation

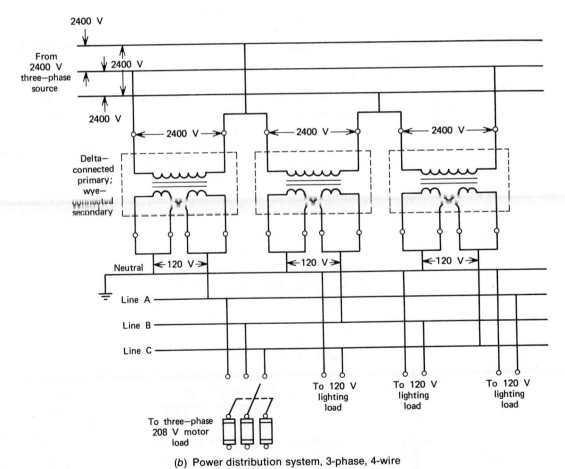

(b) Power distribution system, 3-phase, 4-wire

Figure 12-13 Three-phase transformers and power distribution.

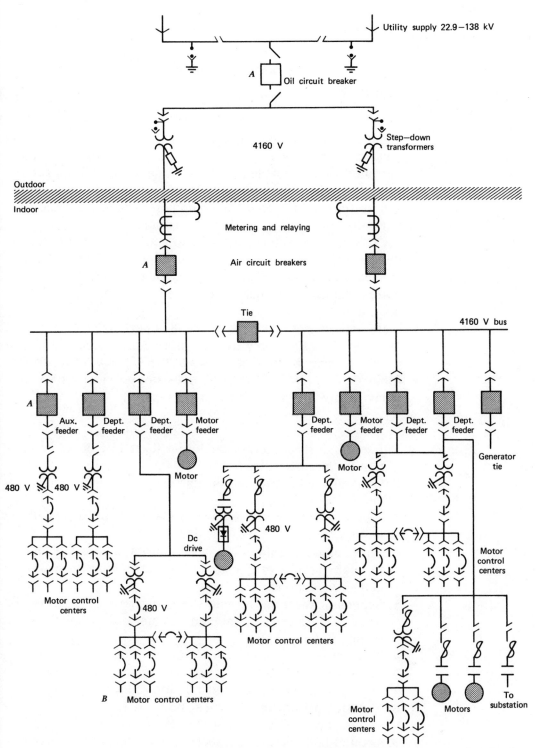

Figure 12-14 Typical power distribution system for industrial plants. (Courtesy of Westinghouse Electric Corp.)

at a lower current and time setting, to effect an isolation if a *fault* persists.

System grounding and *ground fault protection* is required by the National Electrical Code (NEC) for personnel safety, equipment protection, and lightning protection. Grounding conductors must be large enough to carry the maximum ground current for a reasonable time without damage or disconnection of the circuit. Neutral conductors, while grounded at the source, *cannot* be used for equipment grounding since they normally carry unbalanced load currents.

Ground currents may be used to activate ground detecting devices and circuit breakers, as shown at the 3750 kVA transformer (*C*) in Figure 12-15. The 51N coil connected to the 200/5 current transformer operates the circuit breaker in the 13.8 kV main line.

If continuity of service is of prime importance so that a fault may be tolerated only for a short time, then *resistance grounding* may be used as shown at the 3750 kVA transformer (point *C*, Fig. 12-15). The 3.5 ohm resistor provides a 10-second delay before the circuit breaker is operated.

Summary 1. Large industrial motors cannot be started by direct connection to the power lines.
2. Generally industrial motor starting is accomplished at reduced line voltage.
3. Motor speed control is more easily accomplished with dc motors.
4. Three-phase ac motors may be started at reduced voltage through the use of resistors or by switching from wye to delta connections.
5. Schematic diagrams for industrial power systems are usually one-line drawings containing simplified symbols with emphasis on coding.
6. The sequence of power distribution usually proceeds from the top to the bottom of a one-line diagram.
7. High-voltage ac power transmission reduces line losses.
8. Step-down transformers must be used to reduce the high-voltage ac for operation of the lighting and motor load.
9. Ground-fault circuits are necessary for the protection of personnel and equipment against the effects of lightning, static charges, and defective components.

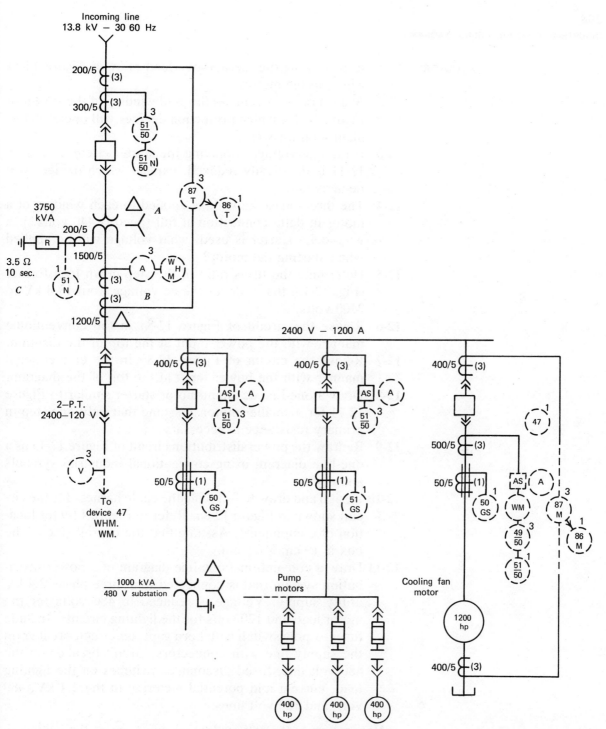

Figure 12-15 Typical control switchboard circuit. (Courtesy of Westinghouse Electric Corp.)

Problems 12-1 Explain how the thermal-overload relay of Figure 12-11 will stop the motor.

12-2 A short circuit occurs in the field winding of the starter in Figure 12-11; which protection devices will operate? Explain your answer.

12-3 If the line voltage supplying the starter circuit in Figure 12-11 is drastically reduced, explain which devices will be activated.

12-4 The three-phase ac voltage applied to each winding of a motor in delta connection at full speed is 440 volts. If a wye–delta starter is used, what voltage will be applied when starting the motor?

12-5 Determine the turns ratio of the step-down transformer (Fig. 12-15) that reduces the ac voltage from 13.8 kV to 2400 volts.

12-6 Revise the circuit of Figure 12-8a in the conventional manner with the power input at the top of the diagram.

12-7 Revise the circuit of Figure 12-8b in the conventional manner with the power input at the top of the diagram.

12-8 Draw a one-line dc shunt-motor starter similar to Figure 12-11 for a smaller motor. Assume that only one step in primary resistance is necessary.

12-9 Redraw the power distribution circuit of Figure 12-13 as a one-line diagram using conventional industrial symbols and codes.

12-10 Design and draw to full scale the cable harness for the circuit shown in Figure 12-8a. Refer to Figure 12-9 for location of components. Assume that the overall size of the box is 19 cm × 25 cm.

12-11 Draw a conventional one-line diagram of a power distribution system that is connected to a three-phase 2.4 kV utility supply. Voltage is reduced to 480 volts for the motor load and 120 volts for the lighting circuits. Include fuses, 3-pole switch with horn gaps, air circuit breaker on the supply line with connectors, circuit breaker on the 480 volt line, fused disconnect switches on the lighting load, current and potential metering in the 2.4 kV, 480 volt, and 120 volt lines.

Now return to the self-evaluation questions at the beginning of this chapter and see how well you can answer them. If you cannot answer certain questions, place a check next to them, and review the appropriate section of the chapter to find the answers.

References **Charles J. Baer,** *Electrical and Electronics Drawing,* Third Edition, McGraw-Hill, New York, 1973, pp. 217–259.

Irving L. Kosow, *Electric Machinery and Transformers,* Prentice-Hall, Englewood Cliffs, N.J., 1972.

Irving L. Kosow, *Control of Electric Machines,* Prentice-Hall, Englewood Cliffs, N.J., 1973.

George Shiers, *Electronic Drafting,* Prentice-Hall, Englewood Cliffs, N.J., 1962, pp. 327–333.

Chapter 13 Graphs and Charts

Instructional Objectives

1. To understand the value of data presentation in graphical form.
2. To become familiar with many methods of graphic presentation.
3. To learn to select the best method of graphic presentation for a particular set of data.
4. To learn how scales are chosen and captioned.
5. To become proficient in plotting data.
6. To prepare properly drawn and easily interpreted graphs.
7. To understand why every graph must be a complete presentation and include all constant factors.
8. To develop an ability to design conversion scales and nomographs.

Self-Evaluation Questions

Test your prior knowledge of the information in this chapter by answering the following questions. Watch for the answers as you read the chapter. Your final evaluation of whether you understand the material is measured by your ability to answer these questions. When you have completed this chapter, return to this section and answer the questions again.

1. Why is graphic presentation of data preferred to data tabulation?
2. What is a dependent variable?
3. What is an independent variable?
4. What factors determine the choice of scales on a graph?
5. How is statistical data presented graphically?
6. On which axis is the dependent variable usually plotted?
7. On which axis is the independent variable usually plotted?
8. When is it necessary to plot graphical data in four quadrants?
9. Where are the independent and dependent variables located on a graph, respectively?
10. What information must be included in the scale captions and how should it be presented?
11. Under what circumstances is a curve drawn from point-to-point?
12. How should an ordinate caption be lettered?
13. Describe data applications that are best presented on *polar coordinate* graph paper.

14. What kind of data requires use of *semilog* graph paper?
15. What information must be included in the title of any graph?
16. Of what value is a nomograph?
17. What is a Smith chart and where is it used?

13-1
Pie and
Bar
Graphs
A *graph* is a visual method of displaying tabulated data for analysis, solving problems, and comparing results. *Graphic representations* are used extensively to present engineering facts, statistics, and physical phenomena. A pictorial or graphic description is much easier to understand than a numerical tabulation or a verbal description.

Statistical data are often presented graphically (Fig. 13-1). The average person usually can rapidly determine the quantitative relationship from a *pie* graph (Fig. 13-1*a*) or a *bar* graph (Fig. 13-1*b*), which presents the *same data* in another form. For example, it is quickly noted in Figure 13-1*b* that less money was spent for ICs in 1970 than for any other electronic component category.

Pie charts are much more difficult to prepare since each pie wedge must be calculated first as a percentage of the whole and then converted to degrees.

13-2
Rectilinear
Graphs
A graph is a diagram that represents variations in the relationship between two or more *variables*. Variables are quantities that may change in value or magnitude under different conditions. When two variables, such as X and Y, are so related that a change in the value of X causes a change in the value of Y, then Y is said to be a *function* of X. By assigning values to X and solving for the value of Y in the general equation $Y = f(X)$, X is then the *independent variable* and Y is the *dependent variable*.

Scales are needed since, by definition, the values of X and Y will change. Usually the independent variable (X) is scaled *horizontally* near the bottom of the graph paper and is known as the *abscissa*. The vertical scale or *ordinate* is always drawn at the left hand edge of the graph paper and is used to plot the dependent variable (Y).

Most graphs are plotted using positive values since most graphical data consists of positive numbers. Figure 13-2 illus-

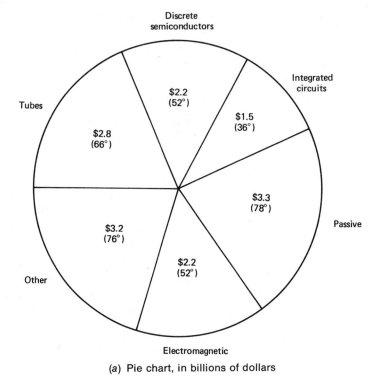

(a) Pie chart, in billions of dollars

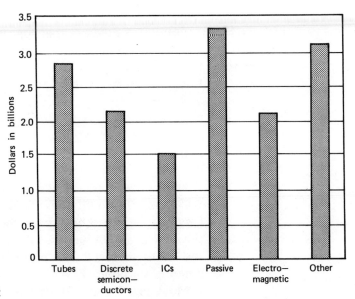

Figure 13-1 Total worldwide component consumption in 1970, $15.2 billion.

(b) Bar chart

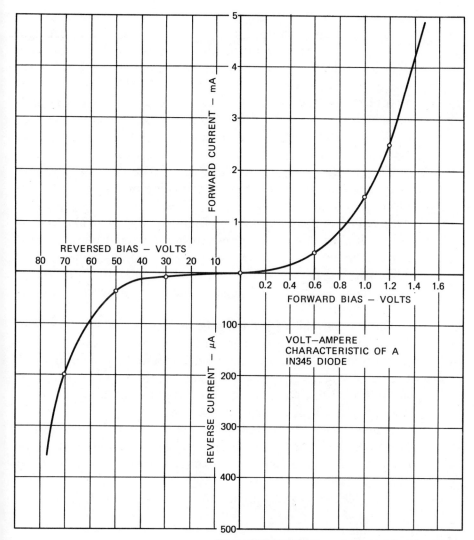

Figure 13-2 Volt-ampere characteristic of a diode.

trates the plot of data which also have negative values. The electrical characteristics of a diode are completely different when the applied potential is reversed, as noted by the change in reversed-biased voltage and reverse current scale.

The *origin* of the graph is at the intersection of the abscissa and ordinate scales and is usually located (Fig. 13-3) in the lower left-hand corner of the graph paper. In Figure 13-2, the *origin* or zero point for all scales has been moved to the *center* of the sheet; the result is a division of four combinations of positive and negative values known as *quadrants*. The upper right or *first quadrant* plot indicates that a positive voltage (independent variable) causes a positive current to flow through the diode. When the bias voltage is made negative, the current is also negative; therefore, the data is plotted in the *third quadrant*. It will be noted that the value per scale division of voltage is different to the left of the origin; similarly, the current scales are not the same. If the scales were identical, values on the graph would be difficult to read. The *choice of scales is therefore very important* to prevent crowding and to simplify the plotting of data.

Specific points are *plotted* by following the vertical and horizontal grid lines corresponding to the collected data. For example, the data in Table 13-1 was plotted in the graph of Figure 13-2. The first pair of values indicate that when the bias voltage (X) is zero, the current (Y) is zero; the point is therefore plotted at the origin.

The second pair of points indicate that when $X = 0.6$ volt, $Y = 400$ μA $= 0.4$ mA. The value 0.6 volt is located on the horizontal or X scale and the 0.4 mA is found on the upper vertical or Y scale; vertical and horizontal grid lines, respectively, are followed to their intersection. The intersection point is recorded as a dot or a very small circle.

Similarly, the remainder of the tabulated data is also plotted. Negative values of voltage (X) are to be located to the left of the origin and negative values of current (Y) are found on the lower vertical scale.

Most electrical phenomena are continuous, and therefore a

Table 13-1
Volt-Ampere Characteristic of a Diode

Bias (volts)	Current (μA)
0	0
0.6	400 (0.4 mA)
1.0	1500 (1.5 mA)
1.2	2500 (2.5 mA)
-30	-10
-50	-40
-70	-200

smooth average curve should be drawn through the plotted points. Do *not* connect the plotted points with straight lines unless the graph represents a straight line relationship between *X* and *Y*.

Occasionally, it is found that a plotted point appears to be completely out of line with preceding and following points; this may result from ineptness of measurement, poor quality instruments, or an error in recording data. When the accuracy of measured data is in doubt, a *smooth* curve should be drawn so that almost the same number of points lie above and below the curve. A clear plastic *French curve* (Fig. 1-8*a*) is useful in locating and drawing this *average* curve.

13-3 Graph Construction Standards Many technical problems can be solved by construction and analysis of a graph. When a graph is carefully drawn with suitable scales and proportions, its accuracy should be of the same order as the accuracy of the instrumentation in the circuit being tested. The following generally accepted standards should be followed.

1. The graph should be free of all lines and lettering that are not essential to the reader's clear understanding of its message.
2. On the other hand, a graph must be complete in itself and contain notes regarding all *constant factors* (Figs. 13-3 and 13-4).
3. Scales should be chosen so as to ensure effective and efficient use of the coordinate area (Fig. 13-4).
4. The horizontal or independent variable scale values usually should increase from *left* to *right* in first quadrant plots, and the vertical or dependent variable values from *bottom* to *top* (Figs. 13-3 and 13-4).
5. The origin should be displaced to the right and upward from the lower left-hand printed corner so that space is available for the scales and captions (Fig. 13-3).
6. Avoid the use of digits at each scale division (Fig. 13-3).
7. Avoid the use of large-scale digits; use larger units of measurement. For example, use kilohms or kΩ instead of ohms (Fig. 13-7).
8. All lettering should be placed so as to be easily read from the bottom and the right-hand side of the graph.

9. The vertical scale caption should be lettered from bottom to top (Fig. 13-2).

10. All lettering should be simple vertical uppercase Gothic.

11. The scale captions should include both the name of the variable and its unit of measurement (Fig. 13-2).

12. Each coordinate ruling should represent 1, 2, or 5 units of measurement or their decimal equivalents when using arithmetical scales (Fig. 13-2).

13. When plotted data points are included, small circles, squares, or triangles may be used to distinguish between points whenever several curves appear on the same graph.

14. Curves should preferably be drawn with *smooth, solid* lines since most physical phenomena are continuous. (point-to-point lines are *only* used on graphs of *discontinuous* data).

15. When the plotted points appear to be discontinuous as a result of difficulty or errors in measurement, the average smooth curve should be drawn so that an equal number of points lie on both sides.

16. Each curve of a family of curves must be identified by lettering directly. A legend should *not* be used (Figs. 13-3 through 13-5).

17. A graph must contain a clear and concise title; it must also contain revelant supporting data (Fig. 13-2).

**13-4
Logarithmic
Graphs**

If one or both of the scale divisions are proportional to the logarithm values of the variable, then the *logarithmic* (log) *scale* can better accommodate a large range of values. A log scale is subdivided according to the common logarithm (base 10) of the uniform divisions as shown in Figure 13-4. *Both* scales of Figure 13-4*b* show one logarithmic *cycle* or *modulus* from 1 to 10; single cycles may have ranges such as 0.1 to 1.0, or 10 to 100, or 100 to 1000, and so forth. The log *cycle* therefore covers some *power of ten*. It is impossible to have a zero on a log scale; the zero is represented graphically because the log of 1 *is* zero!

Another advantage of logarithmic graphs is in comparing the rectilinear, semilog, and log-log graphs (Figs. 13-3 and 13-4). The general form $Y = X^n$ is plotted for three values of n on three different types of graph paper. Note that while an improved display is obtained through use of semilog paper as

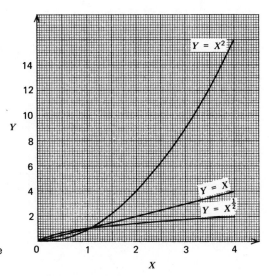

Figure 13-3 Rectilinear graph of the general form $Y = X^n$.

against rectilinear paper, *straight line* plots are obtained for all three curves using *log-log* graph paper (Fig. 13-4b). The graph of Figure 13-4b shows that the steepness or *slope* of the curve is proportional to the exponential power of n. This exponential value (if not previously known) could be obtained from the experimental data when plotted on log-log paper.

Semilog graphs (Fig. 13-4a) are preferred when showing a comparison of *rate of change* rather than an amount of change; the log scale is used as the ordinate or dependent variable.

**13-5
Conversion
Scales**
A given quantity must often be expressed in terms of another system of measurement, such as length in inches converted to centimeters. For this purpose, a simple *conversion scale* is prepared. Figure 13-5 illustrates a conversion scale for finding the *mantissa* of a common logarithm. Similar conversion scales may be drawn to any convenient size, if samples of *both* scales are available, using the following procedure.

1. The decimal scale is marked on line AB to a convenient length.
2. A transfer line AC is drawn at an angle of approximately 45° to the line AB.
3. The log scale is marked on line AC. Almost any length cycle from log graph paper that is available may be used.
4. Line BC is drawn between the extremes of the two scales.

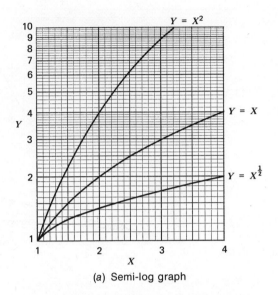

(a) Semi-log graph

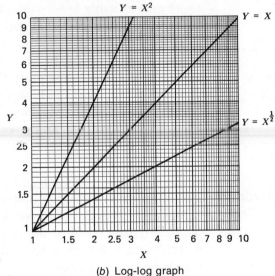

Figure 13-4 Logarithmic graphs of the general form $Y = X^n$.

(b) Log-log graph

5. Lines are drawn parallel to BC from each of the log scale (AC) divisions.

6. The points of intersection of the parallel lines with line AB represent the decimal numbers of the conversion scale.

The accuracy and number of significant digits obtainable is obviously determined largely by the length of the conversion

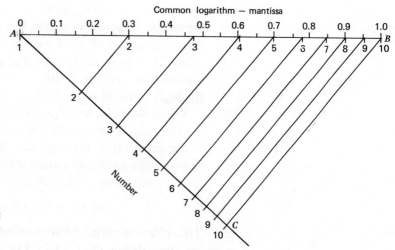

Figure 13-5 Projection of a common log scale and construction of a conversion scale to decimal numbers.

scale and the care in transferring scales. Such conversion scales are sometimes drawn on linear graph paper to improve reading accuracy.

A draftsman sometimes may need an *odd ratio* drafting scale such as 1 cm = 7 m; a *ratio scale* may be similarly prepared, saving the time spent in repetitive calculations.

13-6 Nomographs

Nomograph, or *alignment charts,* are extensions of conversion scales used to derive solutions to specific problems. Conversion scales contain only two variables and are on adjacent scales, while nomographs involve at least three variables and scales arranged in a specific geometrical relationship. The simplicity of using nomographs has won such favor in technology; they are easily and advantageously used by nontechnical personnel as well.

The simplest type of nomograph (Fig. 13-6) consists of three parallel scales so graduated that a straight line joining points on two of the scales will cut the third scale at a point that satisfies the relation between the variables. The expression $X = A + B$ can be solved by a nomograph in which the linear scales of X, A, and B are parallel to each other. The X scale is usually drawn equidistant from the A and B scales. The X scale may be calibrated by drawing straight lines between points on the A and B scales, marking the intersection points on the X scale. The value at the intersection point is obviously the sum of the A

and B values. Each combination may be drawn and calculated; however, after a few points have located, the remaining values may be marked off with a divider. It may be observed that if the X and A values are known and plotted on this nomograph, the value of B may be found by extending the line from point A through point X to an intersection on scale B (Fig. 13-6).

When the solution of a problem involves multiplication or division, nomograph scales are usually logarithmic. For example, if the solution of the equation $X_L = 2\pi f L$ is desired, the equation may be expressed similar to the previous example as an addition nomograph through the use of logarithms:

$$\log X_L = \log (2\pi) + \log f + \log L$$

Therefore, a parallel line nomograph may be drawn if the scales are logarithmic (Fig. 13-7). For example, if a line is drawn from scale A at the 100 kHz point to the 10 mH point on the D scale, the reactance scale C is intersected at approximately 6 kΩ. Al-

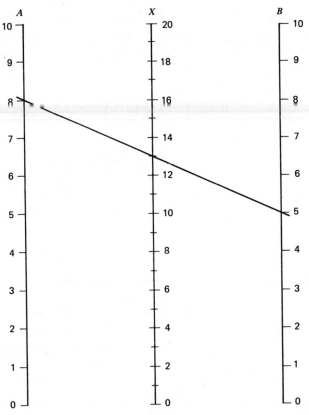

Figure 13-6 Simple addition nomograph.

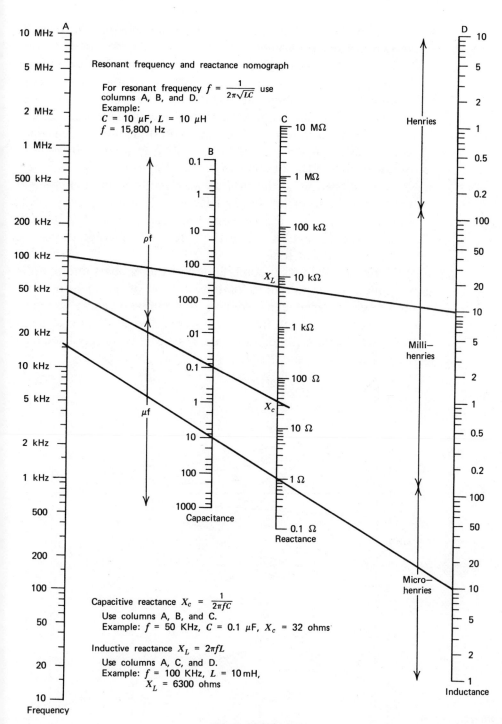

Resonant frequency and reactance nomograph

For resonant frequency $f = \dfrac{1}{2\pi\sqrt{LC}}$ use columns A, B, and D.
Example:
$C = 10\ \mu F,\ L = 10\ \mu H$
$f = 15,800$ Hz

Capacitive reactance $X_c = \dfrac{1}{2\pi fC}$
Use columns A, B, and C.
Example: $f = 50$ KHz, $C = 0.1\ \mu F$, $X_c = 32$ ohms

Inductive reactance $X_L = 2\pi fL$
Use columns A, C, and D.
Example: $f = 100$ KHz, $L = 10$ mH,
$X_L = 6300$ ohms

Figure 13-7 Chart to determine inductive or capacitive reactance, or resonance.

most any length log scale may be drawn using the method shown in Figure 13-5.

The *N chart* (so-called), shown in Figure 13-8, is another form of nomograph used for the solution of an equation containing three variables. If the spacing between the vertical scales is about equal to the length of the scales, the intersection at the diagonal *tie line* is sharper. Note that the vertical lines contain linear scales and that one scale is graduated in ascending order while the other is in descending order.

The calibration of the tie line may be done mathematically using the theorem of similar triangles. Equal accuracy is attained by locating the intersections using typical values as shown in Figure 13-8. Note that the tie line graduations are nonlinear, but *not* logarithmic.

Figure 13-9 shows a combination of the N chart and parallel alignment chart methods for the solution of an equation containing *four* variables. Under such circumstances, it is necessary to obtain the solution in two steps. The quotient *D/d* is not numerically determined at the *axis* scale; it is merely used as a

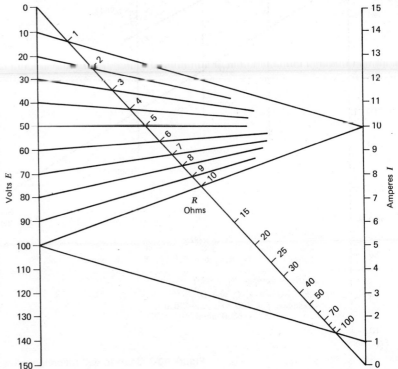

Figure 13-8 *N*-type chart for Ohm's law, *I* = *E/R*.

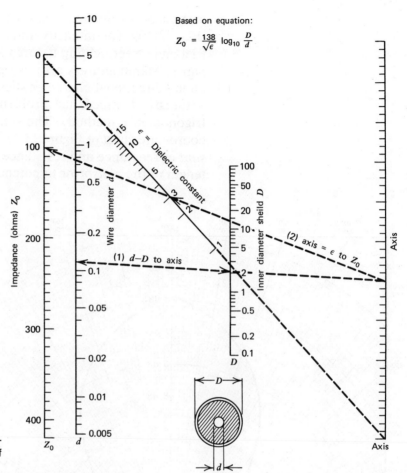

Based on equation:

$$Z_0 = \frac{138}{\sqrt{\epsilon}} \log_{10} \frac{D}{d}$$

Figure 13-9 Nomograph for determination of the characteristic impedance of a coaxial cable.

pivot for the second step in determining characteristic impedance.

13-7 Polar Graphs **13-7** *Polar coordinates* are concerned with the location of a point through the magnitude of the shortest distance from the origin to the point and the angular direction of that point from the origin. The two variables therefore consist of a *linear* measurement and an *angular* measurement. Polar coordinate graph paper (Fig. 13-10) indicates linear magnitude by equally spaced *concentric circles;* the angular quantity is plotted radially with respect to the origin of the circles.

Directional characteristics, such as the plot of the radiation pattern of an antenna, may be plotted on polar graph paper (Fig. 13-10). The intensity and distribution of light from industrial and street lighting fixtures are also plotted on polar graph paper. Manufacturers also use polar graphs to present microphone directional characteristics.

Electrical impedance problems are generally solved using trigonometric methods. The solution is simplified by the polar coordinate chart of Figure 13-11. Since impedance is the *phasor sum* of reactance and resistance, the impedance is equal to the length of the *phasor* or hypotenuse of a triangle, whose base is

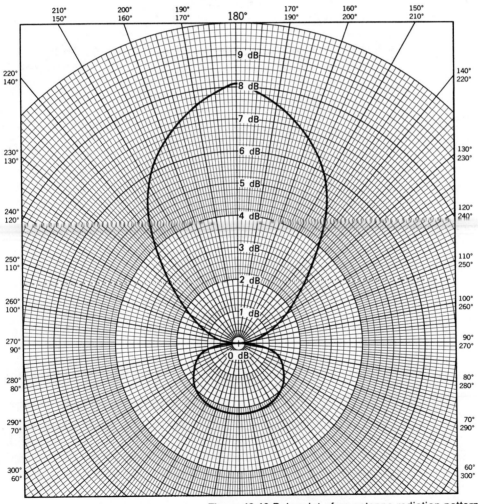

Figure 13-10 Polar plot of an antenna radiation pattern.

resistance and height is reactance. Impedance in polar form may also be obtained from this chart; it is represented by the magnitude of the phasor and the angle between the phasor and the baseline.

13-8
The Smith
Chart

The *Smith chart,* or *polar impedance diagram,* is shown in Figure 13-12. It consists of two sets of circles, and arcs of circles, so arranged that various important quantities connected with mismatched transmission lines may be plotted and evaluated. The *centers* of the complete circles all lie on the horizontal straight line. The circles correspond to various values of *normalized resistance* along the line. The arcs above and below the horizontal line correspond to *normalized line reactance.* The plot in Figure 13-12 indicates a normalized load impedance of $0.6 + j0.3$ ohms.

By striking an arc through point P with center at the chart's center, so that it intersects the horizontal line, a *standing wave*

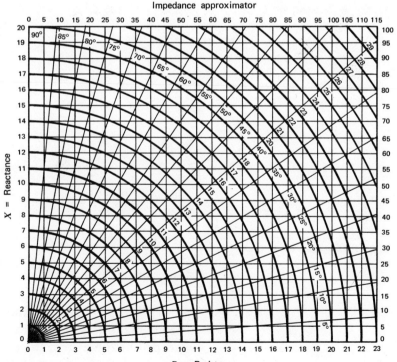

Figure 13-11 Impedance polar chart.

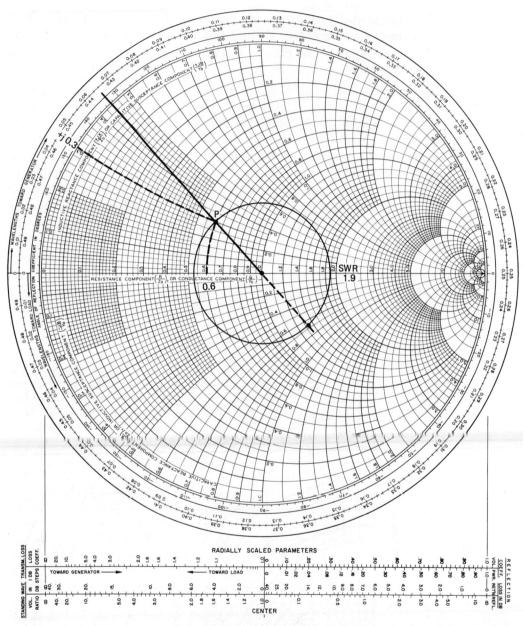

Figure 13-12 Smith chart for determination of the standing wave ratio (SWR).

ratio (SWR) of 1.9 is obtained. Standing waves are produced when some of the total power applied to a transmission line is absorbed by the load at the end of the line; the difference is reflected back to the generator. The ratio of maximum current to minimum current along a transmission line is the SWR. It is therefore a measure of the mismatch between the load and the line. When the load is perfectly matched to the line, the SWR is unity. The Smith chart may also be used to determine the location and length of a short-circuited stub of similar transmission line which will bring the SWR to unity with no loss of transmitted power.

Summary 1. Semipictorial graphs such as pie charts and bar charts summarize statistical data graphically for easier understanding.
2. A set of technical data, when properly presented graphically, facilitates analysis.
3. Line graphs must be carefully prepared to the generally accepted standards of mathematicians and engineers.
4. The drafting accuracy in preparing a graph must equal the accuracy of the instrumentation used to obtain the data.
5. Every graph must be complete in itself; it should not be necessary to refer elsewhere for supporting data.
6. Logarithmic scales should be used when the range of plotted values is large or to yield equations that best represent the data.
7. Polar graphs are used to show the directional characteristics of a device, such as an antenna.
8. When a project involves many similar and tedious conversions of data to a different system of measurement, the drafting of a conversion scale is worth the effort.
9. The geometrical arrangement of graduated lines known as a *nomograph* saves time in the repetitious solution of mathematical equations.

Problems **13-1** In 1970, $11.8 billion was spent on consumer electronic equipment, $20.8 billion on government electronic equipment, and $24.2 billion on industrial electronic equipment. Properly prepare a 10 cm diameter pie chart containing this information.

13-2 Prepare a bar graph for comparison of electrical energy consumption in the following countries: USA, 220 million kW; Great Britain, 50 million kW; Russia, 95 million kW; West Germany, 40 million kW; Japan, 35 million kW.

13-3 Draw a conversion scale for wire diameter in millimeters to mil diameter from the following data:

Mm diameter	2.54	5.08	7.62	10.16	12.70
Mil diameter	100	200	300	400	500

Assume that 254 mm equals 1000 mils.

13-4 Plot a family of curves of the following data in which the voltage is the independent variable.

Volts	6.0	8.0	10	12	14	
Current, mA	3.0	4.0	5.0	6.0	7.0	at constant 2 kΩ
Volts	12	18	24	30	36	
Current, mA	2.0	3.0	4.0	5.0	6.0	at constant 6 kΩ
Volts	16	20	24	28	32	
Current, mA	2.0	2.5	3.0	3.5	4.0	at constant 8 kΩ

13-5 The following static collector test data of a 2N408 transistor was measured at a constant 0.30 mA base current. plot on rectangular coordinate graph paper assuming the collector current to be the dependent variable.

V_{CE}, volts	0.2	0.3	0.5	1.5	2.5	3.5	4.5	5.5
I_C current, mA	22	26	28	29	30	31	32	33

13-6 A 2N408 transistor is connected as a common emitter amplifier to obtain the following outputs at different test frequencies when the input is maintained constant at 0.050 volt rms. Plot on semilog graph paper.

Frequency, Hz	20	50	100	250	1k	10k	50k	200k
Output, volt rms	0.46	0.97	1.62	2.20	2.25	2.16	1.90	1.67

13-7 To determine the ground reflection factors and direction for maximum field strength of an antenna, the following data is to be plotted on polar coordinate graph paper:

Angle, deg.	0	20	50	90	150	160	170	180
Power, dB	1.0	0.9	0.4	0	1.0	2.0	3.0	4.0

Angle, deg.	190	200	210	270	310	340
Power, dB	3.0	2.0	1.0	0	0.4	0.9

13-8 Prepare a parallel line type nomograph for the solution of the electrical power formula $P = IE$. The logarithmic voltage (E) scale should range from 10 volts to 100 volts, while the logarithmic current (I) scale range is from 1 mA to 10 mA.

13-9 Using the same ranges as in Problem 13-8, draw an N-type nomograph for the power formula.

Now return to the self-evaluation questions at the beginning of this chapter and see how well you can answer them. If you cannot answer certain questions, place a check next to them, and review the appropriate sections of the chapter to find the answers.

References Lamont V. Blake, *Transmission Lines and Waveguides,* Wiley, New York, 1969, pp. 76–96.

Thomas E. French, and Charles J. Vierck, *Engineering Drawing,* Tenth Edition, McGraw-Hill, New York, 1963, pp. 449–462.

John D. Lenk, *Handbook of Electronic Charts, Graphs, and Tables,* Prentice-Hall, Englewood Cliffs, N.J., 1970.

A. S. Levens, *Nomography,* Wiley, New York, 1948.

George Shiers, *Electronic Drafting,* Prentice-Hall, Englewood Cliffs N.J., 1962, pp. 478–521.

Appendix A Acronyms and Abbreviations for Electrical Terms

Adjust	ADJ	Common	COM
Alternating current	ac	Complementary metal oxide	
Alternator	ALT	semiconductor	CMOS
Ammeter	AM.	Conductor	COND
Ampere	A	Connection	CONN
Ampere-turn	AT(NI)	Contact	CONT
Amplifier	AR	Continuous	CONT
Amplitude modulation	AM	Continuous wave	CW
Analog to digital	A/D	Control	CONT
Anode	A	Control grid	CG
Antenna	ANT.	Converter	CONV
Armature	ARM.	Coupling	CPLG
Attenuator	AT	Cross connection	XCONN
Audio frequency	AF	Crystal	XTL
Automatic frequency control	AFC	Crystal oscillator	CO
Automatic gain control	AGC	Current transformer	CT
Automatic volume control	AVC	Cutout	CO
		Cycle (frequency)	CY (C)
Ballast	BALL.	Cycles per second	CPS
Band eliminator	BE		
Band pass	BP	Decibel	dB
Base (transistor)	B	Demodulator	DEM
Battery	BT	Detector	DET
Beat frequency	BF	Digital multimeter	DMM
Beat-frequency oscillator	BFO	Digital volt meter	DVM
Binary coded decimal	BCD	Digital volt-ohm-milliammeter	DVOM
Bypass	BYP	Diode	CR
		Direct current	dc
Calibrate	CAL	Direct current volts	V_{dc}
Carrier	CARR	Direct current working volts	DCWV
Cathode	CATH(K)	Disconnect switch	DS
Cathode-ray oscilloscope	CRO	Discriminator	DISCR
Cathode-ray tube	CRT	Double contact	DC
Center tap	CT	Double pole	DP
Circuit	CKT	Double pole, double throw	DPDT
Circuit breaker	CB	Double pole, single throw	DPST
Coaxial cable	COAX.	Doubler	DBLR
Collector (transistor)	C	Double throw	DT
Color code	CC		

Drain (FET)	D	Junction field-effect transistor	J FET
Dynamotor	DYNM	Kilohertz	kHz
Electrolytic	ELECT.	Kilohm	kΩ
Electromagnetic	EM	Kilovolt	kV
Electromotive force	EMF	Kilowatt	kW
Electron coupled oscillator	ECO		
Electronic voltmeter	EVM	Large-scale integration	LSI
Emitter (transistor)	E	Light	LT
Extremely high frequency	EHF	Light-emitting diode	LED
		Limiter	LIM
Facsimile	FAX	Limit switch	LS
Farad	F	Link	LK
Fast operating	FO	Liquid crystal display	LCD
Fast release	FR	Local oscillator	LO
Federal Communication Commission	FCC	Loud speaker	LS
Field-effect transistor	FET	Low frequency	LF
Filament	FIL	Low voltage	LV
Filter	FLT		
Flip-flop circuit	FF	Magnet (-o)	MAG
Frequency	FREQ	Magnetomotive force	MMF
Frequency modulation	FM	Master oscillator	MO
		Medium frequency	MF
Galvanometer	GALV	Mega (10^6)	MEG
Gate (SCR)	G	Megahertz	MHz
Generator	G	Megohm	MΩ
Grid	G	Metal oxide semiconductor	MOS
Ground	GND(GRD)	Metal oxide semiconductor field-effect transistor	MOS FET
Heater	HR (H)	Meter	M
Henry	H	Micro (10^{-6})	μ
Hertz	Hz	Microfarad	μF
High frequency	HF	Microphone	MIKE(MIC)
High pass	HP	Milli (10^{-3})	m
High potential	H POT.	Milliamperes	mA
High voltage	HV	Modulator	MOD
Horizontal (CRT)	H	Momentary contact	MC
		Monitor	MON
Ignitor	IGN	Motor	M
Impulse	IMP	Motor generator	MG
Inductance (-ion)	IND	Multimeter	MULTR
Insulate	INS	Multiplex	MX (MPX)
Insulated gate FET	IGFET	Multipole	MP
Integrated circuit	IC	Multivibrator	MVB
Intercommunication	INTERCOM		
Intermediate frequency	IF	Negative	NEG
Inverter	INV	Network	NET.
		Neutral	NEUT
Jack	J	Normally closed	n.c.
Joint	JT	Normally open	n.o.
Junction box	JB		

Not connected	NC	Reactive volt-ampere	var
Not in contact	NIC	Receiver	RCVR
		Receptacle	RECP
Ohm	Ω	Recorder	REC
Operational amplifier	OP AMP	Rectifier	RECT
Oscillator	OSC	Regulator (-ed)	REG
Overload	OL	Relay	K
		Remote control	RC
Parallel	PAR.	Repeater	REP
Peak-to-peak	p-p	Resistor	RES
Pentode	PENT.	Resistor-transistor logic	RTL
Permanent magnet	PM	Root mean square	RMS
Phase	PH		
Phase modulation	PM	Saturable reactor	SR
Phonograph	PHONO	Screen	SCRN
Pick-up	PU	Screen grid	SG
Pico (10⁻¹²)	p	Secondary	SEC (S)
Picofarad	pF	Selector	SEL
Plate	P	Series	SER
Plug	P	Shield	SHLD
Polarity	PO	Signal	SIG
Polarize	POL	Silicon-controlled rectifier	SCR
Positive	POS	Single contact	SC
Potential	POT.	Single phase	1 PH
Potential difference	PD	Single pole	SP
Power	PWR	Single side band	SSB
Power amplifier	PA	Single throw	ST
Power factor	PF	Slow operate	SO.
Power oscillator	PO	Slow release	SR
Preamplifier	PREAMP	Socket	SOC
Primary	PRI (P)	Solenoid	SOL
Public address system	PA	Source (FET)	S
Printed circuit	PC	Speaker	SPKR
Pulsating current	PC	Supply	SUP
Pulse-amplitude modulation	PAM	Suppressor	SUPPR
Pulse-code modulation	PCM	Switch	SW
Pulse frequency	PF	Switchboard	SWBD
Pulse-position modulation	PPM	Synchronous	SYN
Pulse repetition frequency	PRF		
Pulse repetition rate	PRR	Telegraph	TLG
Pulses per second	PPS	Telemeter	TLM
Pulse-time modulation	PTM	Television	TV
Pulse-width modulation	PWM	Terminal	TERM.
Push button	PB	Terminal board	TB
Push-pull	P-P	Tertiary	TER (T)
		Test link	TL
Radar	RDR	Test point	TP
Radio	RAD	Test switch	TSW
Radio frequency	RF	Thermocouple	TC
Random access memory	RAM		

Thermostat	THERMO	Unijunction transistor	UJT
Three phase	3 PH	Vacuum tube	VT
Three-pole	3P	Vacuum tube voltmeter	VTVM
Time constant	TC (T)	Variable frequency oscillator	VFO
Time delay	TD	Vertical (CRT)	V
Times ten	X10	Very-high frequency	VHF
Total harmonic distortion	THD	Vibrator	VIB
Transceiver	XCVR	Video	VID
Transformer	T	Video frequency	VDF
Transistor	Q (*npn,pnp*)	Voice frequency	VF
Transistor-trans.logic	TTL	Volt	V
Transmission	XMSN	Voltage-controlled oscillator	VCO
Transmit-receive	TR	Voltage regulator	VR
Transmitter	TR	Volt ampere, reactive	var
Transmitting	XMTG	Voltmeter	VM
Trip coil	TC	Volt-ohm-millimeter	VOM
Triple throw	3T	Volume unit	VU
Tuned grid	TG		
Tuned plate	TP	Watt	W
Tuner	TUN	Watt-hour	Wh
		Wattmeter	WM
Ultrahigh frequency	UHF	Wire-wound	WW
Under voltage	UV	Working volts	WV

Appendix B Metric Equivalents Table

Centimeters	Inches	Fraction (inches)	AWG	Current (A)	Centimeters	Inches	Fraction (inches)	Centimeters
0.001	0.0004				0.700	0.2756		
0.002	0.0008				0.714	0.2812	9/32	
0.003	0.0012				0.754	0.2969	19/64	1
0.004	0.0016				0.794	0.3125	5/16	
0.005	0.0020				0.800	0.3149		
0.006	0.0024				0.833	0.3281	21/64	0
0.007	0.0028				0.873	0.3437	11/32	
0.008	0.0032		40		0.900	0.3543		
0.009	0.0035		39		0.913	0.3594	23/64	
0.010	0.0039		38		0.953	0.3750	3/8	00
0.020	0.0078		32		0.992	0.3906	25/64	
0.030	0.0118		29		1.000	0.3937		
0.040	0.0158	1/64	26		1.032	0.4062	13/32	000
0.050	0.0197		24		1.072	0.4219	27/64	
0.060	0.0236		23		1.100	0.4331		
0.070	0.0276			3	1.111	0.4375	7/16	
0.080	0.0315	1/32	20	5	1.151	0.4531	29/64	
0.090	0.354		19		1.191	0.4687	15/32	
0.100	0.0394		18	7	1.200	0.4724		
0.119	0.0469	3/64			1.230	0.4844	31/64	
0.150	0.0025	1/16	16	10	1.270	0.5000	1/2	
0.198	0.0781	5/64	14	15	1.300	0.5118		
0.200	0.0787		12	20	1.310	0.5156	33/64	
0.238	0.0937	3/32			1.349	0.5312	17/32	
0.278	0.1094	7/64	10	30	1.389	0.5469	35/64	
0.300	0.1181		9		1.400	0.5512		
0.318	0.1250	1/8	8		1.429	0.5625	9/16	
0.357	0.1406	9/64			1.468	0.5781	37/64	
0.397	0.1562	5/32	7		1.500	0.5906		
0.400	0.1575				1.508	0.5937	19/32	
0.437	0.1719	11/64	6		1.548	0.6094	39/64	
0.476	0.1875	3/16	5		1.588	0.6250	5/8	
0.500	0.1969				1.600	0.6299		
0.516	0.2031	13/64	4		1.627	0.6406	41/64	
0.556	0.2187	7/32			1.667	0.6562	21/32	
0.595	0.2344	15/64	3		1.700	0.6693		
0.600	0.2362				1.707	0.6719	43/64	
0.635	0.2500	1/4			1.746	0.6875	11/16	
0.675	0.2656	17/64	2		1.786	0.7031	45/64	

Centimeters	Inches	Fraction (inches)	Centimeters	Inches
1.800	0.7087		4.3	1.693
1.826	0.7187	23/32	4.4	1.732
1.865	0.7344	47/64	4.5	1.772
1.900	0.7480		4.6	1.811
1.905	0.7500	3/4	4.7	1.850
1.945	0.7656	49/64	4.8	1.890
1.984	0.7812	25/32	4.9	1.929
2.000	0.7874		5.0	1.969
2.024	0.7969	51/64	5.1	2.008
2.064	0.8125	13/16	5.2	2.047
2.100	0.8268		5.3	2.087
2.103	0.8281	53/64	5.4	2.126
2.143	0.8437	27/32	5.5	2.165
2.183	0.8594	55/64	5.6	2.205
2.200	0.8661		5.7	2.244
2.223	0.8750	7/8	5.8	2.283
2.262	0.8906	57/64	5.9	2.323
2.300	0.9055		6.0	2.362
2.302	0.9062	29/32	6.1	2.402
2.342	0.9219	59/64	6.2	2.441
2.381	0.9375	15/16	6.3	2.480
2.400	0.9449		6.4	2.520
2.421	0.9531	61/64	6.5	2.559
2.461	0.9687	31/32	6.6	2.598
2.500	0.9843	63/64	6.7	2.638
2.540	1.0000		6.8	2.677
2.6	1.024		6.9	2.717
2.7	1.063		7.0	2.756
2.8	1.102		7.2	2.835
2.9	1.142		7.4	2.913
3.0	1.181		7.6	2.992
3.1	1.220		7.8	3.071
3.2	1.260		8.0	3.150
3.3	1.299		8.2	3.228
3.4	1.339		8.4	3.307
3.5	1.378		8.6	3.386
3.6	1.417		8.8	3.465
3.7	1.457		9.0	3.543
3.8	1.496		9.2	3.622
3.9	1.535		9.4	3.701
4.0	1.575		9.6	3.780
4.1	1.614		9.8	3.858
4.2	1.654		10.0	3.937

Appendix C **Resistor Color Code**

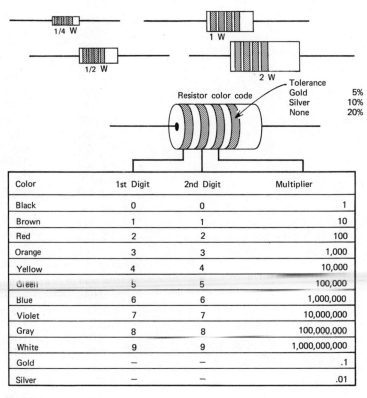

Color	1st Digit	2nd Digit	Multiplier
Black	0	0	1
Brown	1	1	10
Red	2	2	100
Orange	3	3	1,000
Yellow	4	4	10,000
Green	5	5	100,000
Blue	6	6	1,000,000
Violet	7	7	10,000,000
Gray	8	8	100,000,000
White	9	9	1,000,000,000
Gold	—	—	.1
Silver	—	—	.01

Examples:

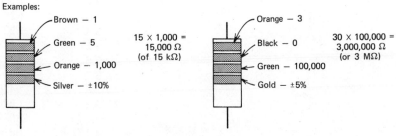

Brown — 1
Green — 5
Orange — 1,000
Silver — ±10%

15 × 1,000 = 15,000 Ω (of 15 kΩ)

Orange — 3
Black — 0
Green — 100,000
Gold — ±5%

30 × 100,000 = 3,000,000 Ω (or 3 MΩ)

Figure A-1 Resistor color code.

Standard
Stock
Resistor
Values

All values are stocked in 5 percent tolerance; available 10 percent tolerance values are indicated with an asterisk.

Ohms	Ohms	Ohms	Ohms	Kilohms	Kilohms	Megohms	Megohms	Megohms
2.7*	16	100*	620	3.9*	24	0.1*	0.62	3.9*
3.0	18*	110	680*	4.3	27*	0.11	0.68*	4.3
3.3*	20	120*	750	4.7*	30	0.12*	0.75	4.7*
3.6	22*	130	820*	5.1	33*	0.13	0.82*	5.1
3.9*	24	150*	910	5.6*	36	0.15*	0.91	5.6*
4.3	27*	160	1000*	6.2	39*	0.16	1.00*	6.2
4.7*	30	180*	1100	6.8*	43	0.18*	1.1	6.8*
5.1	33*	200	1200*	7.5	47*	0.20	1.2*	7.5
5.6*	36	220*	1300	8.2*	51	0.22*	1.3	8.2*
6.2	39*	240	1500*	9.1	56*	0.24	1.5*	9.1
6.8*	43	270*	1600	10.0*	62	0.27*	1.6	10.0*
7.5	47*	300	1800*	11.0	68*	0.30	1.8*	11.0
8.2*	51	330*	2000	12.0*	75	0.33*	2.0	12.0*
9.1	56*	360	2200*	13.0	82*	0.36	2.2*	13.0
10*	62	390*	2400	15.0*	91	0.39*	2.4	15.0*
11	68*	430	2700*	16.0		0.43	2.7*	16.0
12*	75	470*	3000	18.0*		0.47*	3.0	18.0*
13	82*	510	3300*	20.0		0.51	3.3*	20.0
15*	91	560*	3600	22.0*		0.56*	3.6	22.0*

Wiring Color Code

Table D-1
Chassis Wiring

Color	Abbreviation	Code Number	Circuit
Black	BK	0	Grounds, grounded elements
Brown	BR	1	Filament heaters, off ground
Red	R	2	Power supply B-plus
Orange	O	3	Screen grids
Yellow	Y	4	Cathodes or emitters
Green	GN	5	Control grids, base
Blue	BL	6	Plates, anodes, collectors
Violet (purple)	V	7	Power supply B-minus
Gray	GY	8	ac power lines
White	W	9	Miscellaneous, returns above or below ground, AVC, etc.

Table D-2
Power Transformer Leads

Color	Stripe	Winding lead
Black		Primary common, if tapped
Black	Yellow	Primary tap
Black	Red	Primary finish
Red		High voltage plate winding
Red	Yellow	High voltage center tap
Yellow		Rectifier filament winding
Green	Yellow	Rectifier filament winding center tap
Brown		Filament winding No. 2
Brown	Yellow	Filament winding No. 2 center tap
Slate		Filament winding No. 3
Slate	Yellow	Filament winding No. 3 center tap

Table D-3
Audio Transformer Leads

Blue	Plate lead of primary, finish
Red	B-plus lead
Brown	Plate lead on center-tapped primaries, start
Green	Grid lead to secondary, finish
Black	Grid return
Yellow	Grid lead on center-tapped secondary, start

Table D-4
Intermediate Frequency Transformer Leads

Blue	Plate lead
Red	B-plus lead
Green	Grid (or diode) lead
Green, black stripe	Plate, second diode, center-tapped secondary
Black	Grid (or diode) return, secondary center tap

Table D-5
Industrial Control Circuits

Black	Line, load and control circuit at line voltage
Red	ac control circuit
Blue	dc control circuit
Yellow	Interlock panel control when energized from external force
Green	Equipment grounding conductor
White	Grounded neutral conductor

Appendix E Typical Mechanical Data of Semiconductors

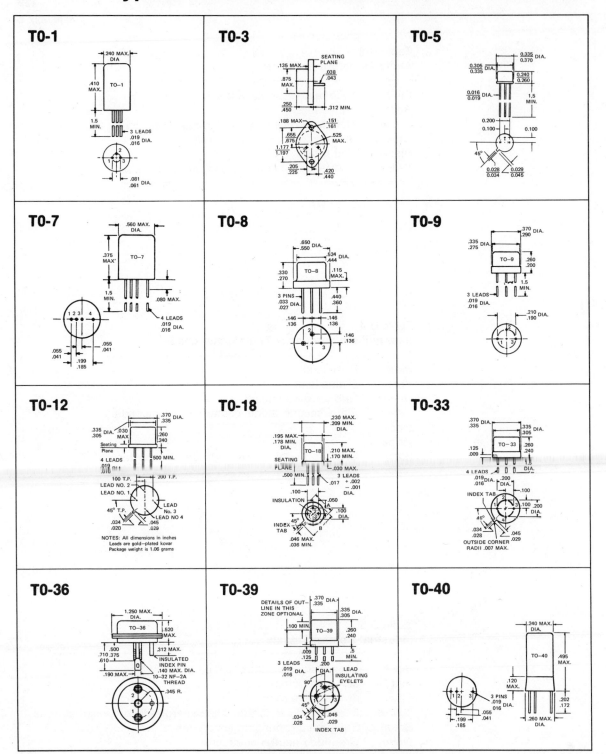

Figure A-2 Typical mechanical data of transistors.

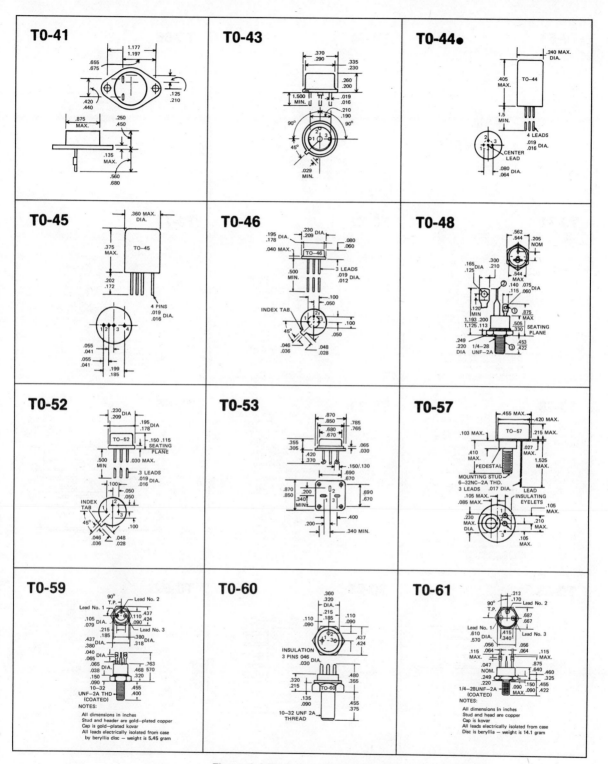

Figure A-3 Typical mechanical data of semiconductors.

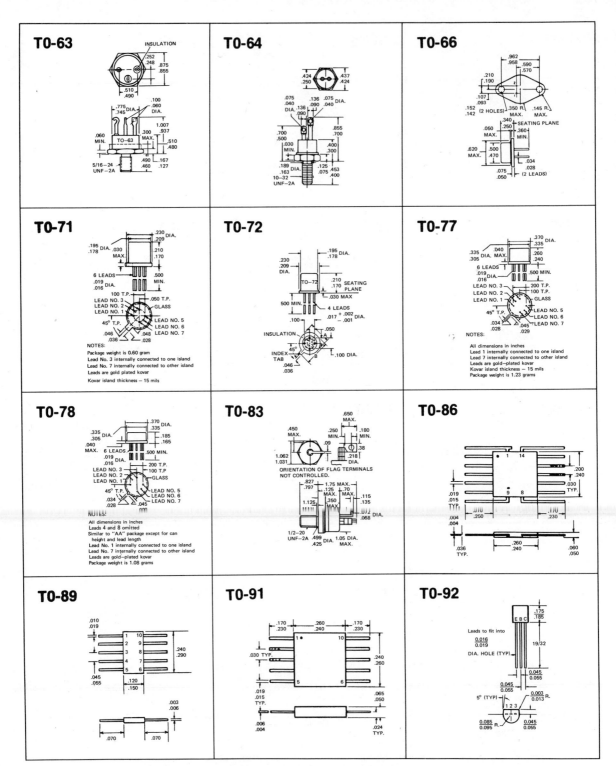

Figure A-4 Typical mechanical data of semiconductors.

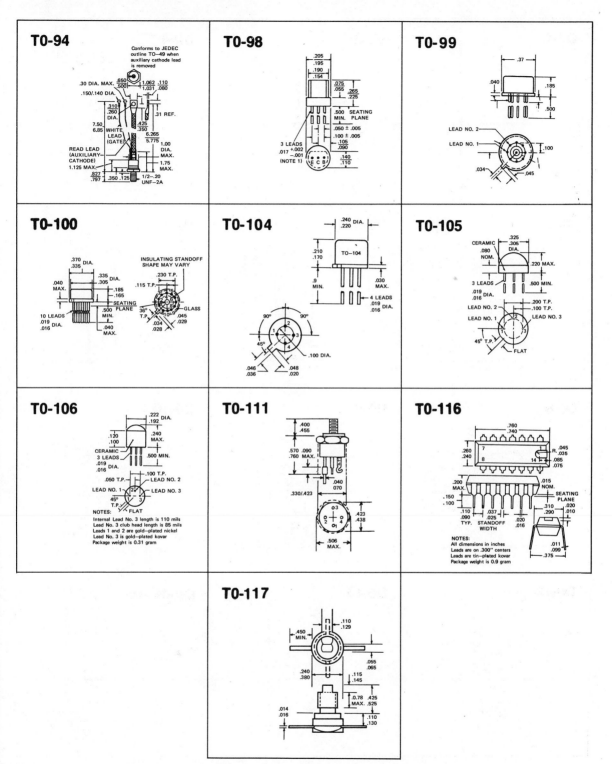

Figure A-5 Typical mechanical data of semiconductors.

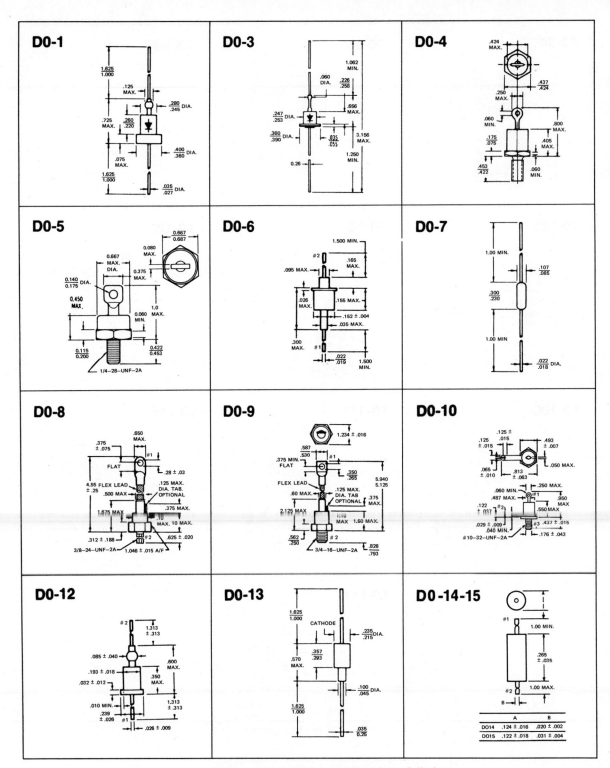

Figure A-6 Typical mechanical data of diodes.

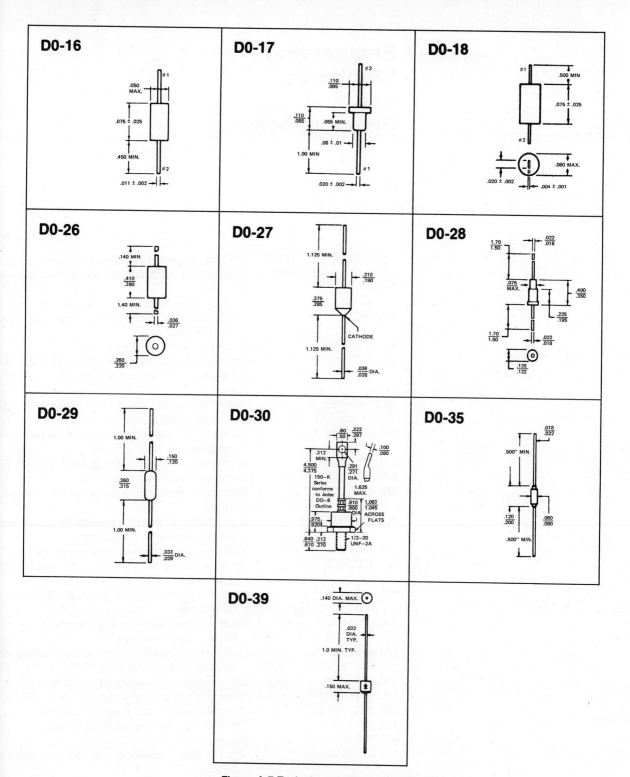

Figure A-7 Typical mechanical data of diodes.

Engineering Standards and Specifications

ANSI Y32.2, IEE 315	Electronic-electrical graphic symbols.
ANSI Y32.9 1972 IEEE	Symbols for architectural layout diagrams.
F-1	**Military**
A.B.M.A. Std. 428	Printed circuit design and construction.
MIL-STD-27	Designations for electric switchgear and control devices.
MIL-STD-100	Engineering drawing practices.
MIL-STD-108	Definitions and basic requirements for enclosures for electric and electronic equipment.
MIL-STD-189	Racks, electrical equipment, 19-in. and associated panels.
MIL-STD-195	Marking of connections for electrical connections.
MIL-STD-196	Joint electronics type designation system.
MIL-STD-221	Color code for resistors.
MIL-STD-242	Electronic equipment parts, selected standards.
MIL-STD-242	Types and definitions of models for communication-electronic equipment.
MIL-STD-275	Printed wiring for electronic equipment.
MIL-STD-283	Letter symbols for electrical and electronic quantities.
MIL-STD-429	Printed wiring and printed circuit terms and definitions.
MIL-STD-681	Identification coding and application of hookup and lead wire.
MIL-STD-686	Identification marking and color coding of electrical cable and cord.
MIL-STD-701	List of standard semiconductor devices.
MIL-STD-710	Military standard, synchros, 60 and 400 Hz.
MIL-STD-1313	Microelectronic terms and definitions.
MIL-P-13949	Plastic sheet, laminated, copper-clad.
MIL-P-55110	Printed wiring boards.
MIL-S-46844	Solder bath soldering of printed circuit boards.
F-2	**Institute of Printed Circuits**
IPC-A-600	Acceptability of printed Circuit Boards.

IPC-CF-150	Copper foil for printed circuit application.
IPC-D-300	Standard tolerances for printed circuit.
IPC-FC-218	Flexible flat cable type connectors.
IPC-FC-240	Flexible printed wiring specifications.
IPC-TC-500	Copper foil for printed circuit application.
IPC-TC-510	Clinched wire type interfacial connections.
IPC-TC-550	Interfacial connections specifications.
IPC-R-700	Rx for repair of printed wiring boards.
IPC-ML series	Multilayer design guide.

Appendix G Relay Operation Codes

AC, or sine symbol	Alternating current
D	Differential
DB	Double biased
DP	Dash pot
EP	Electrically polarized
FO, or small circle	Fast operate
FR, diagonal line	Fast release
MG	Marginal
NB	No bias
NR	Nonreactive
P, or +	Magnetically polarized
SA	Slow operate or slow release
SO, or X	Slow operate
SR, or III	Slow release
SW	Sandwich wound

Appendix H Circuit Board Eyelets

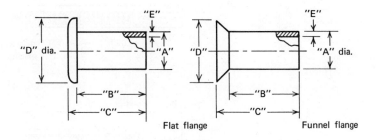

Flat flange   Funnel flange

Board Thickness (mm)	Hole Size (mm)	"A" (mm)	"B" (mm)	"D" (mm)	"E" (mm)	"A" (mm)	"B" (mm)	"D" (mm)	"E" (mm)
0.38	1.32	1.19	1.22	2.03	0.18				
0.79	1.32	1.19	1.57	2.03	0.18	1.19	1.57	1.78	0.15
1.57	1.32	1.19	2.36	2.03	0.18	1.19	2.39	1.78	0.15
2.36	1.32	1.19	3.18	2.03	0.18	1.19	3.18	1.78	0.13
0.79	1.70	1.50	1.57	2.29	0.18	1.50	1.57	2.67	0.18
1.57	1.70	1.50	2.36	2.29	0.18	1.50	2.39	2.67	0.18
2.36	1.70	1.50	3.18	2.29	0.18	1.50	3.18	2.67	0.18
0.79	1.32	1.19	1.57	2.03	0.18				
0.79	1.32	1.19	1.35	2.03	0.18	1.19	1.37	2.03	0.15
1.57	1.32	1.19	2.36	2.03	0.18				
1.57	1.32	1.19	2.16	2.03	0.18	1.19	2.16	2.03	0.15
2.36	1.32	1.19	3.18	2.03	0.18				
2.36	1.32	1.19	3.60	2.03	0.18	1.19	2.97	2.03	0.15
0.79	1.70	1.50	1.57	2.41	0.18	1.50	1.73	2.79	0.18
0.79	1.70	1.50	1.35	2.41	0.18				
1.57	1.70	1.50	2.36	2.41	0.18	1.50	2.51	2.79	0.18
1.57	1.70	1.50	2.16	2.41	0.18				
2.36	1.70	1.50	3.18	2.41	0.18	1.50	3.33	2.79	0.18
2.36	1.70	1.50	3.00	2.41	0.18				
3.18	1.70	1.50	3.96	2.41	0.18	1.50	4.11	2.79	0.18
3.18	1.70	1.50	3.89	2.41	0.18				
Tolerances		±0.051	0.127	0.127		0.076	0.76	0.76	

Appendix I Circuit Board Connectors

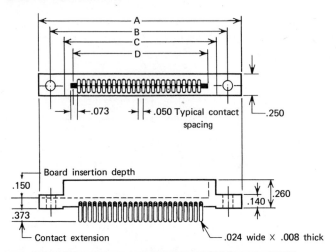

.073

.050 Typical contact spacing

.250

Board insertion depth

.150

.373

Contact extension

.260

.140

.024 wide × .008 thick

Dimensions			
A	B	C	D
2.000	1.750	1.500	1.350

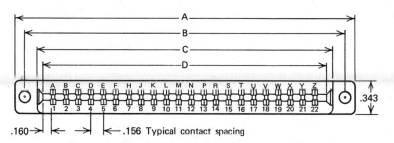

.160

.156 Typical contact spacing

.343

Depth of card insertion — .291 min.

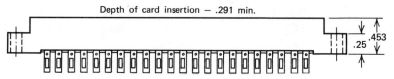

.453

.25

No. of contacts		A	B	C	D	E
Single	Dual					
15	30	3.250	2.937	2.650	2.504	2.687
18	36	3.718	3.406	3.118	2.972	3.156
22	44	4.343	4.031	3.742	3.596	3.781

Contact Spacing	Terminal Rows	Contacts per Row	Fig. No.	Contact Rating	Operating Voltage	Max. Contact Resistance
.050″	2	25	1	1A	350 V DC	.015 ohm
.156″	1	15	2	5A	450 V DC	.006 ohm
.156″	2	15	2	5A	450 V DC	.006 ohm
.156″	1	18	2	5A	450 V DC	.006 ohm
.156″	2	18	2	5A	450 V DC	.006 ohm
.156″	1	22	2	5A	450 V DC	.006 ohm
.156″	2	22	2	5A	450 V DC	.006 ohm

Appendix J Standard Drafting Paper and Film Sizes

A	8.5 × 11 in.
B	11 × 17 in.
C	17 × 22 in.
D	22 × 34 in.
	28 × 40 in.
	34 × 44 in.
A6	10.5 × 14.8 cm
A5	14.8 × 21.0 cm
A4	21.0 × 29.7 cm
A3	29.7 × 42.0 cm
A2	42.0 × 59.4 cm
A1	59.4 × 84.1 cm
A0	84.1 × 118.9 cm

Glossary

Abscissa. The horizontal distance of any point in a coordinate system from the ordinate or Y-axis (Fig. 13-2).

Acoustic device. A device pertaining to the act or sense of hearing (Fig. 4-11c).

Acronym. A word formed by combining initial letters or syllables and letters of a series of words or a compound term.

Active device. An electronic component containing voltage and/or current sources such as transistors (Fig. 4-8).

Airline wiring diagram. A connection drawing in which a single horizontal *or* vertical line represents the cable location, with feed lines branching off (Fig. 7-3).

Alternating current (ac). A periodic current that reverses at regularly recurring intervals of time and that has alternately positive and negative values.

Alumina. Aluminum oxide—an insulator.

Amplifier. A device that enables an input signal to control power and capable of an output generally greater than the input signal (Fig. 5-2).

Amplitude modulation (AM). The amplitude of the RF carrier wave varies with the AF signal (Fig. 5-6a).

Anode. Electrode having a positive charge capable of attracting negative charges.

Appliqués. Printed material cut from a sheet and fastened to the surface of another sheet (Fig. 8-2).

Armature. Rotating member of a dc motor or generator.

Armature, relay. The moving element of a relay.

Ascender. The part of a lowercase letter that reaches into the top of the body of the type, as in b, d, etc. (Fig. 2-6).

Attenuation. To reduce in value.

Automatic gain control (AGC). Constant IF and/or RF amplifier gain achieved by varying the emitter or collector current in accordance with the amplitude of the incoming signal (Fig. 5-5).

Band-pass filter. A circuit that conducts only a specific range of frequencies (Fig. 5-7*g* and *h*).

Band-stop filter. A circuit that will not conduct a specific range of frequencies (Fig. 5-7*i* and *j*).

Bar chart. A semipictorial graph in which statistical values are indicated by the length of bars (Fig. 13-1*b*).

Baseline. The bottom guideline of all letters except for the *g*, *j*, *p*,*q*, and *y*.

Baseline connection diagram. Same as airline wiring diagram.

Base, transistor. A region that lies between an emitter and a collector comparable to the grid of a triode.

Bill of material. A list of all materials and components required in the fabrication of an assembly (Fig. 2-8).

Binary number system. A system of numbers with base 2 such that any number can be expressed by 0 or 1 or a combination of these digits.

Bipolar junction transistor (BJT). Semiconductor device in which both majority and minority carriers take part to provide current gain (Fig. 4-8*c*).

Blade. The straightedge part of a T-square.

Block out. To fill in areas so as to prevent movement of resist through the silk screen.

Blowout coil. Extinguishes circuit breaker arc through effect of its magnetic field (Fig. 12-4*e*).

Bond paper. Inferior quality, all-rag drafting paper.

Boolean algebra. A form of algebra originally applied to mathematical analysis of logic and recently applied in the design of digital computers (Fig. 3-8).

Bow compass. A pair of small compasses having no joint but a curved metal strip between the legs, for drawing circles or arcs of small radius.

Branch instruction. To go off from the main operational steps of a computer program.

Branch-off point. Location at which a feed line leaves the cable (Fig. 7-3).

Bread-boarded circuit. The experimental assembly and wiring of a circuit on a flat surface.

Burnishing. Polishing by friction.

Cabled. Two or more insulated conductors are combined within a sheath or laced together.

Cable harness (cableform). Wires of all types and colors are combined into a cable with branching off feed lines.

Call-outs. Location and identification of assembly components using arrows from outside the drawing to the components (Fig. 10-10).

Capacitor. A device consisting of two electrodes separated by a dielectric (Fig. 4-2).

Capital letter. The form of a letter used at the beginning of a sentence, with proper names, and so forth.

Cap line. The uppermost lettering guide line (Fig. 2-6).

Cartridge fuse. A circuit protective device contained in a fiber tube with contacts at each end (Fig. 4-9*m*).

Cathode. Source of electrons in a vacuum tube; the emitter in a diode or SCR.

Ceramic material. Articles made of fired and baked clay.

Channel, FET. A strip of semiconductor material with end terminals called the *source* and *drain*.

Chassis. The framework to which electronic components are attached.

Chemical energy. In an electric cell, the more chemically active electrode loses electrons and becomes positively charged.

Chokes. Inductors designed to impede the current in a circuit over a specified frequency range (Fig. 4-1*a* and *c*).

Chopper circuit. A circuit designed to convert dc or low frequency input to an ac signal of higher frequency.

Circuit breaker. A device to open a circuit automatically on a predetermined overload of current (Fig. 4-9*p*).

Circular mil. Unit of measurement for determining the area of a wire in cross section, equal to the area of a circle having a diameter of 1 mil (0.001 in.).

Clamp contact, resistor. A metallic ring that may be clamped on a wire-wound resistor to serve as a tap (Fig. 4-3*c*).

Coaxial cable. A transmission line having one sheathlike path completely surrounding the other conducting paths (Fig. 4-9*b*).

Collector, transistor. A region through which the primary flow of charge carriers leaves the base.

Color-coded resistor. A system of colors adopted for identification (Appendix C).

Colpitts oscillator. Regenerative (+) feedback is achieved through a split capacitor across the coil (Fig. 5-4*c*).

Common-base circuit (CB). A bipolar transistor amplifier in which the emitter is the input terminal while the output is obtained from the collector (Fig. 5-2*c*).

Common-collector circuit (CC). A bipolar transistor amplifier in which the base is the input terminal while the output is obtained from the emitter (Fig. 5-2*b*).

Common connection. Electrical connection to a conductor common to several circuits such as a ground or battery bus.

Common-emitter circuit (CE). A bipolar transistor amplifier in which the base is the input terminal while the output is obtained from the collector (Fig. 5-2*a*).

Compass, drafting. An instrument for drawing circles or arcs.

Compound field winding. A combination of series and shunt magnetic field windings.

Concentric circles. Circles having a common center.

Conduit. Thin-walled metallic tubing for containment and protection of insulated conductors.

Constant factors. Any characteristic of a circuit that remains the same throughout a set of measurements.

Contactor. Heavy-duty relay whose contacts open or close a primary circuit when its coil, in a second-

ary circuit, is energized (Fig. 12-4).

Conventional current flow. Current flow to a point of negative charge *from* a point of less negative or positive charge.

Conversion scale. A graphical method of changing one system of measurements to another (Fig. 13-5).

Coordinate. Any of a set of magnitudes by means of which the position of a point, line, or angle is determined with reference to fixed elements (Fig. 13-2).

Coupling. The association of two or more inductors or circuits in such a way that power or signal information may be transferred from one to another (Fig. 5-3*a*).

Crossover, schematic. When two conductors pass over each other usually at 90°, but are not connected to each other.

Cross-sectional area. Area of a plane section of an object at right angles to its length.

Crystal microphone. A high impedance device to convert sound to pulsating dc using the piezoelectric effect (Fig. 4-10*h*).

Current-carrying capacity. The National Electrical Code's recommendation of the maximum current which a conductor may carry without overheating (Appendix B).

Current gain. The ratio of the amount of current flowing out of a circuit to the amount of current flowing into a circuit.

Current transformer. An instrument transformer used to sample the amount of current flowing in a high voltage conductor.

Cycle. One complete successive set of positive and negative values of an alternating current.

Deck, rotary switch. Plastic wafers containing switch contacts and terminals (Fig. 4-6*d*).

Decoder-driver IC. A circuit that changes a coded representation of a number into an amplified decoded version (Fig. 5-8*c*).

Delta connection. A triangular-shaped circuit connection of three windings (Fig. 12-13*a*).

Demand factor. Ratio of the maximum amount of

power required by a system to the total connected load of the system.

Demodulator (detector). A circuit that rectifies a high-frequency wave containing AF intelligence, removes the high-frequency component, and allows only an AF output (Fig. 5-6).

Dependent variable. A quantity whose value is a function of the independent variable (Fig. 13-2).

Descender. The part of a lower case letter that extends below the bottom of most letters, as in g, j, p, and so forth.

Detector circuit. Rectifies a high frequency wave containing AF intelligence and removes the HF component (Fig. 5-6a).

Detent, switch. A mechanical device that either stops or releases the switch position.

Diac. A two-terminal, three-layer, two-junction semiconductor used primarily for triggering triac devices (Fig. 4-7e).

Dielectric loss. Energy not stored in a capacitor but converted to heat or electrostatic field loss at edges.

Diffusion. Movement of impurity materials from a region of high concentration to regions of lower concentration.

Digital readout. A device that indicates the output of a calculator or measuring system in Arabic numeral symbols (Fig. 4-11h).

Dimetric projection. Representation of a three dimensional object on a flat surface in which two scales of measurement and two angles of projection are used (Fig. 10-6).

Diode. An electronic device having an anode and cathode through which current will only flow in one direction (Fig. 4-7a).

Direct coupling. Coupling between amplifier stages without the use of capacitors or transformers (Fig. 5-3b).

Disconnect block. A pullout type insulated housing containing contactors and cartridge fuses (Fig. 11-7).

Discrete components. Electronic parts separate or disconnected from others.

Distribution panel. A control center which distributes electrical energy to branch circuits.

Donuts. Small disks of acid-resisting pressure sensitive material.

Dopant. Impurities such as arsenic or boron added to purified silicon to make semiconductors.

Doping, semiconductor. The addition of impurities to purified silicon or germanium to achieve desired characteristics.

Double-pole switch. Capable of opening or closing both sides of a circuit or two separate circuits (Fig. 4-5b).

Double-throw switch. Capable of opening one circuit while closing another circuit (Fig. 4-5b).

Drafting machine. A parallel-rule mechanism that replaces the T-square, triangle, and protractor (Fig. 1-5).

Drain, FET. One end of an FET channel; function compares to that of the bipolar transistor's collector.

Draw to scale. A drawing made to the same dimensions as the components.

Dropline. The bottom guideline for lowercase letters having descenders, as in g, j, p, and so forth (Fig. 2-6).

Dual in-line package (DIP). A one-to-four circuit IC package having two rows of short stiff terminal pins.

Electrochemical process. Process in which electrical energy is changed to chemical energy or vice versa.

Electrode. Conductor through which an electric current enters or leaves an electrolyte, gas, or vacuum.

Electron. An elementary particle having a negative electrical change.

Electron current flow. Current flow to a point of positive charge from a point of less positive or negative charge.

Electronic voltmeter (EVM). A sensitive, nonloading voltage measuring device.

Ellipse. A plane curve such that the sum of the distances from any part of the curve to two fixed points is a constant (Fig. 10-3).

Emitter, transistor. A region from which charge carriers are injected into the base.

Entrance cable. Fiber-insulated cable of two large-size insulated wires covered with a bare spiral-wound sheath which serves as the neutral conductor (Fig. 11-7).

Entrance head. A weatherproof housing for supporting the unsheathed entrance cable (Fig. 1-2*b*).

Erasing shield. A small rectangle of thin metal or plastic perforated with slots used to protect the lines near those being erased.

Evaporation techniques. Material is vaporized and deposited on a substrate within a vacuum chamber.

Expansion cracks. Transverse breaks in metallic material (on PCB) occasioned by heat expansion.

Feedback. The return of part of the output of a system back to the input or some preceding stage for modification, and control of the output (Fig. 5-4).

Feeder. A set of conductors originating at a main distribution center and supplying power to subdistribution centers or branch circuits.

Feeder material schedules. Tabulation containing the sizes and quantities of conduit, outlet boxes, and hardware.

Feed lines. Lines drawn from a baseline or highway cable to components (Fig. 7-4).

Female connector. Electrical part having a bore or slot designed to receive a correlated inserted part (Fig. 4-9*j*).

Ferrite core. A ceramic material having ferromagnetic properties and used in high frequency transformers to reduce losses (Fig. 4-1*b*).

Field-effect transistor (FET). Three terminal *unipolar* semiconductor in which only the majority carriers are of importance (Fig. 4-8*f*).

Field-loss contactor. A heavy duty relay whose normally-closed contacts open when field current is lost or unduly reduced.

Field rheostat. A variable resistance used to control the current flow through a motor or generator field winding.

Filament, vacuum tube. A tungsten wire or ribbon that is heated to incandescence to produce electrons by passing a current through it.

Filter. A circuit which conducts only predetermined desired frequencies (Fig. 5-7), and rejects unwanted frequencies.

Filter choke. A series-connected inductance with low reactance at the filter frequency (Fig. 5-1*c*).

Flowchart. A schematic diagram showing a sequence of operations (Fig. 3-1).

Flow lines. The path and direction of a sequence of operations in a flowchart (Fig. 3-1).

Foreshortened. To shorten parts of an object's representation to give an illusion of depth while retaining proper proportions (Fig. 10-6).

Freehand. Drawn or sketched by hand without help of rulers or instruments.

Frequency modulation (FM). The AF modulation varies the frequency of the RF carrier wave (Fig. 5-6*c*).

Frequency response. The voltage gain of an amplifier over a range of audio frequencies.

Front view. A drawing of an object as it ordinarily appears when viewed at a distance so that the lines of sight are parallel (Fig. 10-2).

Full-wave rectification. Conversion of both positive and negative portions of an ac wave to dc (Fig. 5-1*b*).

Function. A quantity whose value is dependent on the value of some other quantity (Fig. 13-4).

Function table. A tabulation of rotary switch terminal positions and their circuit relationships (Fig. 6-2*f*).

Fuse. A protective device in which excessive current melts the fuse element and clears a circuit.

Gate, FET. Voltage applied to its gate controls the current flow in the channel of an FET (Fig. 4-8).

Graphite. A soft, black, chemically inert variety of carbon used in making pencils.

Grid, vacuum tube. A screen of wire mesh located between the cathode and plate, effective in controlling electron flow (Fig. 4-6*b*).

Ground. Point of lowest potential which may be a

metallic chassis or common bus (Fig. 5-2).

Ground-fault protection. A leak in current to ground due to defective insulation of components or conductors (Fig. 12-15).

Grown crystal ingot. A crystal bar obtained when a seed crystal is rotated in molten silicon or germanium.

Guidelines. Lines that determine the heights of letters and numbers (Fig. 2-6).

Half-wave rectification. Conversion of either the positive or negative portion of an ac wave to dc (Fig. 5-1a).

Hartley oscillator. An oscillator circuit containing a tapped inductor for purposes of positive feedback (Fig. 5-4a).

Head, T-square. The short vertical member of a T-square that is pressed firmly against the left working edge of the drawing board (Fig. 1-2a).

Heater, vacuum tube. A tungsten filament which, when connected to a low-voltage source, indirectly heats the cathode.

Heat sinks. Metallic surface or device used to absorb and radiate heat.

Helical. A shape produced if a coil is wound in a single layer around a cylinder.

henry. The amount of inductance that allows 1 volt to be induced in a coil when the current changes at the rate of 1 ampere per second.

High-pass filter. A circuit which passes only desired, predetermined high frequencies (Figs. 5-7d, e, and f).

Highway-connection diagram. Conductor lines are merged into long bold horizontal *and* vertical highways; short feed lines branch off from the main cable (Fig. 7-4).

Home run. Cables that return directly to the entrance service box or distribution panel (Fig. 11-6).

Horizontal. A line, plane, or member parallel with the horizon.

Horn gap. Divergent high-voltage switch contacts (Fig. 12-2e).

Hot wire. The black or red insulated coductor that must be fused (Fig. 11-8).

Hydrofluoric acid. An inorganic compound capable of etching glass and ceramic materials.

Hypotenuse. The side of a right triangle opposite the right or 90° angle.

Impedance matching. When the impedance of a generator equals the impedance of its load, then maximum power will be transferred to the load.

Inclined lettering. Lettering that leans to the right at an angle of 67.5° (Fig. 2-2).

Included angle. The angle between two intersecting lines.

Independent variable. The variable to which values are assigned arbitrarily (Fig. 13-2).

Inductance. The ability of a conductor to produce induced voltage when the current varies through the conductor.

Inductor. May consist of a single conductor or a coil of many turns.

In-line terminals. Rotary switch terminals aligned in the shaft direction (Fig. 4-5d).

Input impedance. The opposition to ac into a circuit or device.

Integrated circuit (IC). An electronic circuit which has been fabricated with extremely small components as an inseparable assembly (Fig. 9-7).

Interlock circuit. A circuit that remains activatcd after the original parallel circuit is deactivated (Fig. 12-7).

Intermediate frequency (IF). A difference frequency resulting from mixing the RF carrier and the local oscillator frequency in a superheterodyne receiver.

Interstage transformer. A magnetic coupling device used between amplifier stages (Fig. 5-3c).

Inverter gate. An electronic circuit that reverses the logic level; 1 becomes 0, and 0 is changed to 1 (Fig. 3-8c).

Isometric projection. Representation of a three-dimensional object on a flat surface in which all edges are true dimensions (Fig. 10-1).

Jig. A mechanical device for holding, forming, or processing material.

Jumper wire. Short conductors used between adjacent or almost adjacent terminals.

Jump instruction. An order to skip a portion of a computer's operational steps.

Junction, transistor. The very small region between the emitter and base, and between the base and collector.

Klystron. A microwave oscillator tube in which the electrons are velocity modulated in such a way as to reinforce the tendency to oscillate (Fig. 4-6h).

Ladder diagram. Wiring diagram in which power utilization proceeds from top to bottom (Fig. 12-11).

Laminated-iron core. Made of silicon sheet steel stampings to reduce eddy current losses while increasing magnetic field strength.

Lands. Small circular or rectangular areas of copper foil for soldering of leads (Fig. 8-2).

Latch-down mechanism. A mechanical system used with a set of switches permitting only one to be actuated at a time.

Layout. The overall pattern of a line or schematic diagram.

Legible. Easy to read or decipher.

Lettering template. A plastic strip containing cut-outs of letters and numbers (Fig. 2-9).

Light-emitting diode (LED). A two-terminal semiconductor that emits light when current is passed through it (Fig. 4-7h).

Light-sensitive material. Chemical mixtures that will undergo a physical change when exposed to light.

Linear scale. A scale calibrated with divisions of uniform length.

Lines of sight. Straight lines proceeding from the eye to an object.

Load centers. An electrical power distribution panel.

Logarithm. The exponent of the power to which a given number (base) must be raised in order to equal the quantity.

Logarithmic scale. A nonlinear scale calibrated with divisions proportional to the log of the number (Fig. 13-5b).

Logic blocks. The basic circuits of a computer which combine and manipulate the binary signals so as to perform most of the functions of a computer.

Logic levels. Voltages to represent binary 0 and 1 such as 0 and +5 volts or −3 and +3 volts.

Log-log graph paper. Graph paper in which both the vertical and horizontal scales are logarithmic (Fig. 13-5b).

Lowercase letters. The small letters of the alphabet.

Low-pass filter. A circuit that passes only desired, predetermined low frequencies (Fig. 5-7a, b, and c).

Magnetic coupling. Signal transfer from one circuit to another through electromagnetic induction (Fig. 5-10e).

Magnetron. An ultrahigh-frequency oscillator tube in which a magnetic field is perpendicular to the electric field between cathode and anode (Fig. 4-6i).

Major diameter. The longest distance across an ellipse; length of its major axis (Fig. 10-3).

Male connector. Electrical part, containing pins, designed to be inserted into correlated slots or bores.

Mantissa. The decimal part of a logarithm.

Metallized ink. A suspension of finely divided metallic particles.

Microelectronics. Extremely small components and circuit assemblies made by thin-film, thick-film, or semiconductor techniques.

Microfarads (μF). One millionth of a farad or 1×10^{-6} F; a unit of capacitance.

Microfont lettering. An open style of lettering designed for better photographic reproduction (Fig. 2-3).

Microhenry (μH). One millionth of a henry or 1×10^{-6} H; a unit of inductance.

Millihenry (mH). One thousandth of a henry or 1×10^{-3} H; a unit of inductance.

Mixer circuit. The stage of a superheterodyne receiver in which the input carrier wave is combined with the local oscillator output (Fig. 5-5).

Modulated carrier. A transmitted radio wave containing low-frequency intelligence.

Molex strip contacts. IC female contactors with DIP spacing to be directly soldered into a PCB.

Momentary switch. A switch that is *only* either closed or open when mechanically depressed or actuated (Fig. 4-5c).

Monolithic IC. Electronic circuit fabricated as an inseparable single structure.

Multiturn potentiometer. A potentiometer that contains a helical resistance element, usually of ten turns, permitting greater accuracy of adjustment (Fig. 4-3g).

Multivibrator. Two RC-coupled amplifiers, alternately conducting, having a sawtooth or square wave output (Fig. 5-4e).

N-chart. A nomograph with scales arranged in the shape of the letter N (Fig. 13-8).

Negative logic. The opposite of positive logic levels.

Negative print. Having lights and darks reversed from their original position (Fig. 8-7a).

NEMA. National Electrical Manufacturer's Association.

Neutral wire. A grounded conductor with white or natural gray insulation. It should never be fused or disconnected (Fig. 11-7).

Nibs. The projecting pointed part of a ruling pen.

Nichrome. An alloy of nickel and chrome; resistance wire.

Nomograph. A graph with graduated scales for three or more interrelated variables, so arranged that a straight line joining two values will cut and determine the third variable (Fig. 13-9).

Nonloading. The effect of not reducing actual voltage measurements because of the voltmeter's high internal resistance.

Nonmetallic sheathed cable. Single or multiple conductors within a fibrous protective covering.

Nonpolarized. In nonelectrolytic capacitors, either terminal may be connected to a positive voltage.

Nonshorting. Adjacent rotary switch contacts that are not bridged momentarily by the moving common element (Fig. 4-5d).

Normalize. To divide actual resistance or reactance by the line's characteristic impedance (Fig. 13-11).

Normally closed contact (n.c.). Relay contacts that are closed when the coil is de-energized (Fig. 12-4g).

Normally open contacts (n.o.). Relay contacts that are open when its coil is de-energized (Fig. 12-4f).

Oblique projection. Representation of a three-dimensional object on a flat surface in which the front view has true shape and dimensions (Fig. 10-5).

Offset printing. The inked impression from a lithographic plate is transferred to a rubber-coated cylinder, and then onto the copper-coated board.

One-point perspective. Representing a three-dimensional object on a flat surface to convey the impression of depth and distance using one vanishing point (Fig. 10-7).

Opaque. Impervious to light; not transparent or translucent.

Operational amplifier (OP AMP). A high gain amplifier consisting of a differential amplifier stage followed by a single-ended output stage (Fig. 5-8a).

Operational sequence. The order in which operations are performed (Fig. 3-1).

Optoelectronic relay. A signal applied to the input is light-coupled to the photoconductive cell output.

Ordinate. The vertical distance of any point in a coordinate system from the abscissa or X-axis (Fig. 13-2).

Origin. The point at which the axes of a rectangular coordinate system intersect (Fig. 13-2).

Orthographic drawing. A drawing that includes separate right angle-projected views in addition to the front view (Fig. 1-11).

Oscillator. A circuit capable of converting dc to ac at a frequency and waveshape determined by circuit components (Fig. 4-11*i*).

Oscilloscope. An electronic instrument for projecting the forms of electromagnetic waves on the screen of a cathode-ray tube.

Out-of-phase. Alternating current waves in the same circuit having the same frequency but which do not reach maximum amplitude at the same instant.

Output transformer. A transformer following the final amplifier stage whose secondary impedance matches the load (Fig. 5-3*a*).

Overload relay (OL). A control relay whose contacts will open when the load current becomes excessive (Fig. 12-7).

Parallel. Straight lines or planes that do not intersect, however far extended.

Parts list. A tabulation of all components used in the assembly of a product (Fig. 2-8).

Peak voltage. Maximum amplitude reached by an ac wave.

Perspective drawing. Representing three-dimensional objects on a flat surface to convey the impression of depth and distance (Fig. 10-7).

Phase discriminator. A demodulator stage of an FM superheterodyne receiver (Fig. 5-6*b*).

Phasor. A quantity represented by the magnitude of a line and its direction (Fig. 13-11).

Photoconductive. Resistance of the cell decreases as the intensity of incident light increases.

Photodiode. A two-terminal semiconductor sensitive to light (Fig. 4-7*g*).

Photographic pictorial. A retouched photograph of an assembly containing call-outs for component location and identification (Fig. 10-10).

Photosensitive emulsion. A liquid mixture that is light sensitive.

Photovoltaic. Output voltage of a cell increases with increasing intensity of the incident light (Fig. 4-7*g*).

Picofarads (pF). Formerly called a micro-microfarad; equal to 1×10^{-12} farad. A unit of capacitance.

Picture plane. A visual representation of an object upon a flat surface.

Pie chart. A semipictorial graph in which statistical values are indicated by sectors of a circle (Fig. 13-1*a*).

Pilot lamp. A small lamp used to indicate power is supplied to a device (Fig. 4-11*f*).

Pin location, connector. Connector pins are arranged or numbered so that the location of a specific pin may be found (Fig. 7-1).

Planar transistor. Emitter, base, and collector regions are on the same surface of the device.

Plate, vacuum tube. The anode electrode that has the highest positive potential.

Plug fuses. A circuit protective device contained in an Edison-base screw-in plug.

Point-to-point wiring diagram. Discrete conductors are drawn from origin to destination terminals (Fig. 7-2).

Polar coordinates. A system whereby a point is located by its linear distance from an origin and by the angle from the line to the polar axis (Fig. 13-10).

Polarity. The condition of a body possessing either a positive or negative charge with respect to ground level.

Polarized. In an electrolytic capacitor, the red terminal may only be connected to the most positive potential.

Polyester film. A transparent synthetic sheet material of high tensile strength and dimensional stability.

Polyphase. Having or producing more than one phase.

Positive print. Having the lights and darks in their original relation (Fig. 8-7*b*).

Potential transformer. An instrument transformer used to sample the voltage in a high voltage circuit.

Potentiometer. A three-terminal variable resistance usually connected across a voltage source to vary voltage division (Fig. 4-3e).

Power-handling capability. The ability of an electrical device, such as a resistor, transformer, and the like, to dissipate heat without destroying itself.

Power of ten. The exponent or number that indicates how many times 10 is to be taken as a factor.

Power supply. A circuit that may convert line voltage ac to both dc and ac for operation of electrical equipment (Fig. 5-1).

Primary resistor. Resistors in the input power line to a motor for reduced voltage starting (Fig. 12-8a).

Primary winding. Input winding of a transformer (Fig. 12-13).

Printed circuit. Flat copper conductors bonded to the surface of a flat insulating material.

Projection. A system of parallel lines serving to locate a view at right angles to the front view (Fig. 1-11).

Projectors. A system of parallel lines drawn on a fixed plane representing other visible planes point for point.

Proportional divider. A double-ended divider in which the distance between one pair of points may be adjusted in proportion to the distance between the other pair of points (Fig. 1-9).

Prototype. An original model on which subsequent forms are to be based.

Protractor. A flat, semicircular or circular device engraved in degrees from zero through 180° or 360° (Fig. 1-3).

Pulse generator. An oscillator capable of producing a variety of wave-shaped ac or pulsating dc.

Push-pull amplifier. A circuit in which alternate transistors are cut off during a portion of the input cycle; even harmonics are cancelled (Fig. 5-5).

Quadrant. Any of the four sections formed by the intersection of the X- and Y-axes (Fig. 13-2).

Quadrille paper. Paper printed with uniformly spaced horizontal and vertical lines without margins (Fig. 13-2).

Quartz crystal. A thin section of pure quartz, accurately ground and polished for use in an HF oscillator.

Raceways. A rigid channel or duct of metal or insulating material for containment and protection of conductors.

Rate of change. The amount by which one quantity varies with respect to another quantity.

Ratio detector. An FM demodulator circuit in which one of the two diodes is reversed (Fig. 5-6c).

Reactance. Opposition to ac due to an inductor and/or a capacitor.

Reference marks. Short lines intersecting at 90° for correct location of all dimensions (Fig. 8-9).

Regenerative feedback (+). The in-phase return of part of a circuit's output to its input (Figs. 5-4a, b, and c).

Relaxation oscillator. A circuit producing a sawtooth or pulse output through charge and discharge of a capacitor (Figs. 5-4e and f).

Relay. An electromechanical device that opens or closes an electrical circuit (Fig. 4-1g).

Resist. A paint or liquid that will prevent an acid from etching copper (Fig. 8-1).

Resistance. Opposition to the flow of an electric current.

Resistance-capacitive coupling (RC). Alternating currents are transferred from one stage to another through a network of capacitors and resistors (Fig. 5-3a).

Resonant frequency. The frequency at which inductive reactance cancels capacitive reactance in an LC circuit.

Rheostat. A variable resistance with two terminals usually connected in series in a circuit to vary the current (Fig. 4-3f).

Ribbon cable assembly. Several conductors insulated and molded into a flat cable (Fig. 7-1d).

Right-angled triangle. A drafting instrument in the shape of a triangle in which there is one 90° interior angle.

Ripple voltage. The amount of pulsating dc appearing in an improperly filtered power supply output (Fig. 5-1*b*).

Riser diagram. Shows how electrical energy is distributed through a building from the entrance to the branch load centers.

Rotary switch. A switch whose wiper blade or brush may be moved over an arc of contacts (Fig. 6-2).

Rotor. Rotating member of an ac motor or generator.

Routing chart. A tabulation of all wires in a cable harness including specifications, origins, and destinations of each wire (Fig. 7-2).

Ruling pen. An inking pen in which the ink is stored and released from a pair of flattened points (Fig. 1-7).

Scale, triangular. A three-edged ruler calibrated in six different proportions of length units (Fig. 1-2*d*).

Schematic diagram. An elementary diagram showing the functions and relations of electronic components in a circuit by means of graphical symbols.

Secondary winding. Output winding of a transformer (Fig. 12-13*b*).

Semiconductor. A material with resistivity in the range between metals and insulators.

Semilog paper. Graph paper in which one scale is linear while the other is logarithmic (Fig 13-4*a*).

Sequence of operations. The order in which operations are performed (Fig. 3-1).

Series field. Winding connected in series with a motor or generator armature to produce a magnetic field (Fig. 12-5*f*).

Shielding. Material that substantially reduces the effect of electric or magnetic fields (Fig. 4-1*e*).

Shunt field. Winding connected in parallel with a motor or generator armature to produce a magnetic field (Fig. 12-5*g*).

Silica. White or colorless, crystalline silicon dioxide.

Silicon-controlled rectifier (SCR). A three-terminal semiconductor that conducts as a diode when sufficient positive gate current is applied (Fig. 4-8*a*).

Silicon dioxide. An inorganic compound with superior insulating properties.

Single-pole switch. Capable of opening and closing one side of a circuit (Fig. 4-5*a*).

Single-throw switch. Capable of opening and closing one circuit (Fig. 4-5*a*).

Skip instruction. A program order to pass from one operational procedure to another, omitting what lies between.

Sleeving. A tube of insulating material used for grouping and protection of conductors.

Slide switch. A switch with a projecting post whose movement causes a sliding contact to be made (Fig. 4-5*b*).

Slope. The rise or fall of a line divided by the equivalent horizontal distance.

Slurry. A thin mixture of a liquid with a finely divided substance.

Smith chart. A special polar diagram used to determine the parameters of an electronic transmission line (Fig. 13-12).

Solar cell. A photovoltaic two-terminal electronic device.

Source, FET. One end of an FET channel. Function compares to that of the bipolar transistor's emitter.

Squeegee. An implement with a straight crosspiece edged with rubber used for forcing resist material through a silk screen (Fig. 8-7*c*).

Stages. Part of a circuit including a tube, transistor, or IC and associated components.

Standing wave ratio (SWR). The ratio of maximum current to minimum current along a transmission line (Fig. 13-12).

Stator. Stationary member of an ac motor or generator.

Step-down transformer. A transformer that will reduce the applied ac voltage (Fig. 12-13*b*).

Stranded wire. A number of wires twisted into a single conductor.

Stripping. To remove the insulation from a conductor.

Subassembly. An assembled set of wired devices that is, in turn, part of a larger assembly (Fig. 10-10).

Substrate. Physical material on which an IC is fabricated.

Switch. A device used to open or close a circuit.

Tapped choke. A special inductor in which connection may be made at a point within the winding.

Template, drafting. A flat, clear plastic sheet containing precision cutouts with various shapes and contours.

Terminal lugs. Small tinned metal strips or posts to which a conductor is soldered.

Terminal pads. Rectangular strips of copper foil on a PCB for soldering of leads (Fig. 8-2i).

Test points. Critical circuit points at which a voltmeter or oscilloscope is connected to verify normal operating characteristics (Fig. 5-1b).

Tetrode. A four-element electron tube (Fig. 4-6c).

Thick-film. Deposition of passive circuit elements and interconnections on a ceramic substrate using the silk-screen process.

Thin-film. Evaporation and deposition of circuit elements and interconnections on a silicon substrate.

Three-phase circuit. Having or producing three ac waves displaced in time from each other by one-third of a cycle (Fig. 12-13).

Thyrister. A bistable semiconductor device having three or more junctions that can be switched ON or OFF such as an SCR or triac.

Tickler coil. A winding that will induce a portion of an amplifier circuit output into the input winding (Fig. 5-4b).

Tie line. A calibrated line diagonally connecting two parallel scales of an N-chart (Fig. 13-8).

Titanium dioxide. An inorganic compound with excellent insulating properties.

Title box. A box in the lower right-hand corner of a drawing containing the name of the part or assembly, and other pertinent information (Fig. 2-8).

Toggle-type switch. A device with a projecting lever whose movement causes contact to be made with a snap action (Fig. 4-5a).

Tolerance. Amount by which the measured value can be different from the standard value, expressed as a percent (Appendix C).

Transducer. A device that will convert one form of energy into another form (Fig. 4-11).

Transformer. A device consisting of one or more coils for introducing mutual coupling between electrical circuits (Fig. 4-1f).

Transformer coupling. Alternating current signals are transferred from one amplifier stage to another by induction (Fig. 5-3c).

Transistor. Semiconductor device with three or more terminals capable of the transfer and amplification of a signal (Fig. 4-8c).

Translucent. Allowing the passage of light, but not permitting a clear view of objects beyond.

Transparent. Admitting the passage of light and permitting a clear view of objects beyond.

Travelers. Hot black or red conductors in a 3-wire circuit controlled by 3-way switches (Fig. 11-8).

Triac. Three-terminal semiconductor which conducts as a diode when sufficient gate current is applied of either polarity (Fig. 4-8b).

Triangles, drafting. Clear plastic triangles containing 30°, 60°, and 90° or 45°, 45°, and 90° interior angles.

Trigger circuit. Provides a portion of the input signal so as to synchronize an oscillator's output frequency.

Trimmer capacitor. A small mica dielectric type capacitor which provides small, final adjustments of capacitance (Fig. 4-2d).

Truth table. Used in symbolic logic, in which the "truth" or "falsity" of a statement is listed for all conditions (Fig. 3-8).

T-square. A drafting instrument used to draw parallel horizontal lines (Fig. 1-2a).

Tube-point pen. A specialized ink pen in which a precise line width is obtained (Fig. 1-8).

Tunnel diode. A two-terminal semiconductor that may be used as an amplifier, oscillator, or switching device (Fig. 4-7c).

Turns ratio. The ratio of the number of secondary winding turns to primary turns of a transformer.

Two-gang tuning capacitor. Two variable capacitors with a common rotor (Fig. 4-2c).

Two-point perspective. Representing three dimensional objects on a flat surface to convey an impression of depth and distance using two vanishing points (Fig. 10-9).

Vanishing point. In perspective drawing, the point at which parallel lines appear to converge (Fig. 10-7).

Varactor. A two-terminal semiconductor device in which its capacitance is voltage dependent (Fig. 4-7d).

Variable. A quantity which may change in value under different conditions.

Vellum, drafting. A rag paper that has been treated to give it greater transparency.

Voltage regulator. A device that maintains the output voltage of a power supply independent of load within limits (Fig. 6-6).

Waistline. The intermediate guideline for lowercase letters (Fig. 2-6).

Watt (W). A unit of electrical power equal to the product of the in-phase current drawn by a device and the voltage applied across the device.

Waveform drawings. Simplified drawings of oscilloscope patterns obtained at the test points (Fig. 5-1b).

Wye connection. Connection of at least three windings in a circuit like a Y or three-pointed star (Fig. 12-13a).

X-ray tube. A high vacuum diode which emits x rays when a very high accelerating voltage exists between cathode and anode (Fig. 4-6j)

Zener diode. A pn junction diode reverse-biased into the breakdown region; used for voltage regulation (Fig. 4-7b).

Index